Donald Magderson
7 Meikle Gardens
Westhill
Aberdeenshire

Glasgow Kelvin Grove
3rd November 2023
Uncle David's Funeral

ISLAMIC MAPS

قصر السلسله
حمام عمران

ISLAMIC MAPS

Yossef Rapoport

Bodleian Library
UNIVERSITY OF OXFORD

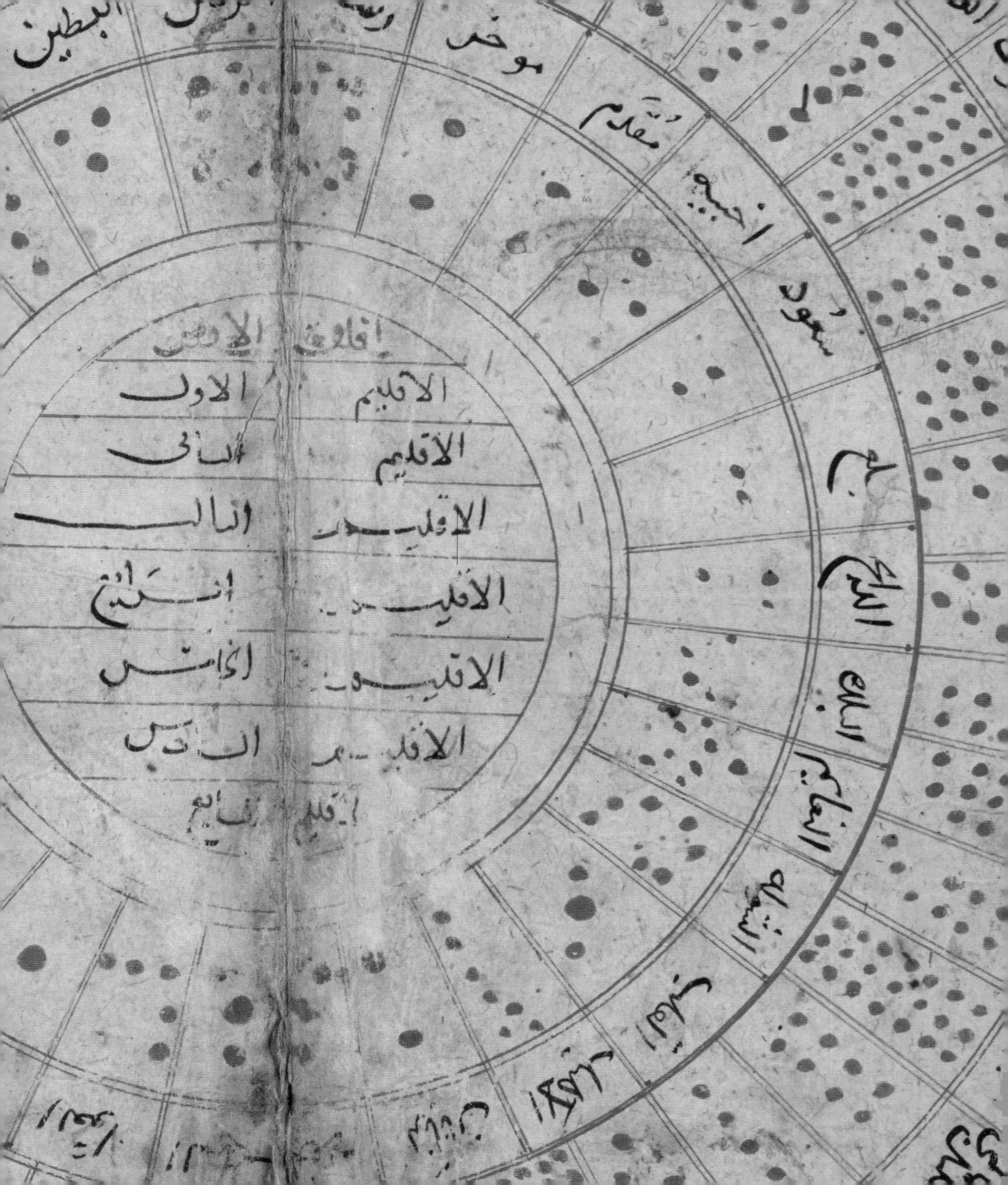
البطين
موخر
مقدم
سعود
النعايم
الاقليم الاول
الاقليم الثاني
الاقليم الثالث
الاقليم الرابع
الاقليم الخامس
الاقليم السادس
اقليم السابع

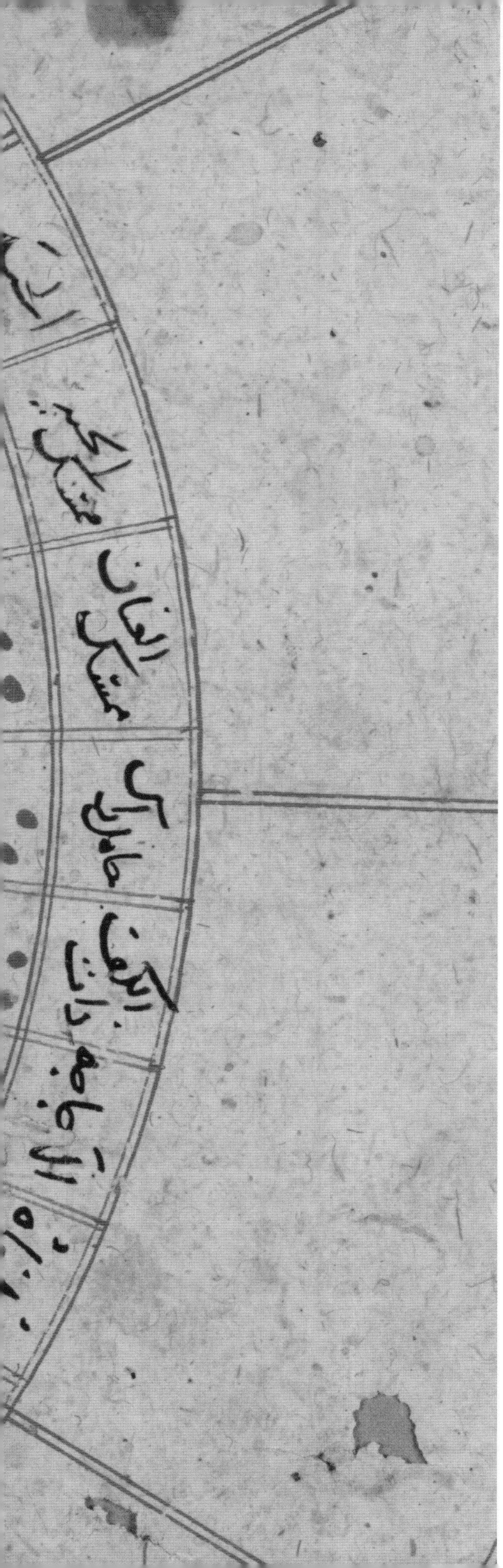

Contents

Introduction

ISLAMIC SOCIETIES PRODUCED SOME of the most extraordinary and captivating maps known to us from the pre-modern world, maps that reflected the unique traits and remarkable achievements of Islamic civilization. Located at the centre of the Old World, Muslims were better placed than anyone else to explore the edges of the inhabited world, and their maps stretched to the far horizons of their geographical knowledge. For most of the Middle Ages, Muslim scholars were at the forefront of mathematical and astronomical sciences, and they were able to apply this knowledge to their representation of physical space. At the same time, Muslim artists had developed distinctive styles, often based on geometrical patterns and calligraphy. Muslim map-makers then brought together that geographical knowledge, science and beauty, and added to it novel cartographical approaches and concepts. The results could be aesthetically stunning, mathematically sophisticated and politically charged, as well as a celebration of human diversity. Perhaps no other type of artefact captures so many dimensions of Islamic civilization, no other object is a better window into the world views of Islamic societies.

This book tells the story of these maps and of the men who made them. Although a few of the maps are familiar, they are commonly deployed as decoration, adorning book covers and illustrating general surveys of Islamic history, without any investigation of their meanings and arguments. On the one hand, the specialized study of Islamic maps has largely been taken up by historians of science, whose main focus has traditionally been mathematical precision and methods of projection, often at the expense of the social and political context in which the maps were produced. On the other hand, in general surveys of the history of cartography Islamic maps are relegated to the margins. Paralleling common views of Islamic civilization, Islamic maps make their appearance only as a prelude to what is generally seen as the more successful and precise early modern and modern European mapping.

In recent decades, and in particular since the onset of the digital age, Islamic maps have attracted much more attention than before.[1] Partly, this is a consequence of a sea change in their visibility and accessibility. Online resources now allow scholars and the general public to view, compare and analyse a previously unimaginable range of maps, which until not long ago were buried in library archives and seen only by a select group of determined and well-funded academics. The increasing fascination with Islamic maps is also part of a growing interest in the history of maps more widely, fuelled by new mobile technologies that have revolutionized the way in which we locate ourselves and find our way in the world. And, of course, ours is also a time in which our notions of Islam itself are being contested. Against a one-dimensional view of Islam that is derived solely from legal texts, the study of material objects brings to the fore a more complex, more pluralistic image of Islamic societies that is more faithful to the realities of the past. In this struggle over the future direction of Islam, maps have a major part to play.

In order to understand the maps produced in pre-modern Islamic societies, we must first shed some of our modern expectations of maps. In particular, we will miss much of what these maps have to say if we judge them solely by their accurate representation of physical space. Mathematical precision is only one facet of Islamic mapping, not always the most exciting one and often not the prevailing consideration of the map-makers themselves. In the past, the modern obsession with mathematical precision has badly distorted the meaning of Islamic maps. Some have dismissed the images discussed in this book as mere paintings, not really 'maps', by implication disparaging the map-makers who made them. Others have romanticized the extent of the mathematical precision of these maps, taking them out of their historical context and employing them as part of a futile competition with the West in which Islam always has to come first.

Instead, this book views Islamic maps as 'a series of ingenious arguments'.[2] Since the ninth century, Muslims have been mapping themselves and others as a way of making sense of the world: of its physical space, of its political boundaries and of its religious, communal and regional identities. Their maps not only reflected the world they lived in, but also shaped the way in which they and others saw the world. They are ultimately acts of interpretation. Maps interpret spaces, and create new ones; maps select what to highlight and what to suppress. Our definition of what qualifies as a map is therefore expansive and is not limited to variations of mathematical geography. Following Harley and Woodward's famous approach, the maps discussed in this book include any 'graphic representations that facilitate a spatial understanding of things'.[3] The term pre-modern Muslims used for map was *ṣūrah*, meaning simply 'image', often in conjunction with *arḍ*, meaning 'Earth' – for them a map was an image of the world, equivalent to the Latin *imago mundi*. Like artists, map-makers were makers of images, and their images, like works of art, offered a multilayered perspective on the world in which they lived.

I employ here a broad definition of maps, and my definition of 'Islam' is likewise expansive, not limited to religious beliefs and ritual practices. The maps that are discussed here were all made by men who identified themselves as Muslims. The labels on all these maps were written in Arabic script: medieval examples were written in the Arabic language, while early modern maps used Arabic letters for Turkish and Persian. Map-makers constructed their worlds with reference to traditions of learning and repositories of knowledge that circulated in the cultural world of Islam, and that included maps made by earlier generations of Muslim map-makers. Their intended audiences were almost always Muslims, although the most famous Muslim map-maker al-Idrīsī produced his maps in the Norman Christian court of Sicily.

Islamic maps were not, generally speaking, religious artefacts. They were guides to this world, not to salvation. Even maps showing the prayer direction towards Mecca, the subject of the final chapter of this book, do not generally carry theological statements. In the background, traditionalist objections to the making of any image (as incompatible with strict Islamic norms) may have have tainted the production of maps. Maps are rarely if ever found in overtly religious texts, such as commentaries on the Qu'ran or legal texts. With the exception of early modern instruments for finding the direction to Mecca and representations of the Kaaba, maps were not displayed in mosques or in other religious structures. This marked a difference from late-medieval European practice, where world maps were hung on the walls of churches and monasteries. As a result, nearly all the extant Islamic maps are part of literary manuscripts, not standalone pieces.

This book examines Islamic visual interpretations of the world in their historical context and in their own terms, from the ninth to the seventeenth centuries. The focus here is on the map-makers themselves: what was the purpose of their maps? what choices did they make? what was the argument they were trying to convey? The map-makers under discussion reflect the diversity of the Islamic world. Most are Sunnis, but some are distinctly Shia; some worked for caliphs, sultans and shahs, while others produced their maps for a market of merchants, scholars and sailors. They hail from Isfahan in the east to Palermo in the west, from Istanbul in the north to Cairo and Aden in the south. Many of the maps they made have come to us through manuscript copies of the original. The objects we see today are thus the products of copyists and illustrators of later generations, who were prone to adapt the original image in light of their own concerns. Maps, by their very nature, are collaborative projects. None of these maps represents Islam as a whole; none of them captures the entire tradition. Moreover, all of them were inspired by cartographic influences outside the world of Islam, whether Greek, Persian or European. As far back as the historical eye can see, there is no purely Islamic map, only hybrid maps that show the wonderful synthetic ability of Islamic civilization.

One of the earliest geographical texts written in the Islamic world, the ninth-century *Routes and Realms* by Ibn Khurradādhbih, contains an anecdote about a map with magical powers. According to a report attributed to the lord of the town of Falluja in Iraq, the local ruler had at his disposal a three-dimensional model representing his entire estates and their irrigation network. Whenever his peasants did not pay their taxes, he would block the canals of these peasants on the model, causing the real canals to dry up; alternatively, he could open the dams on the model, and the corresponding lands in the real world would flood. Thus, the map conferred on him magical powers over the space it represented.[4] This account has been suppressed in later Islamic texts, probably because it has such distinctly pagan overtones. But the magic of maps never disappeared. The maps constructed by Muslim map-makers not only reflected their world but also shaped it, together with its modern legacy.

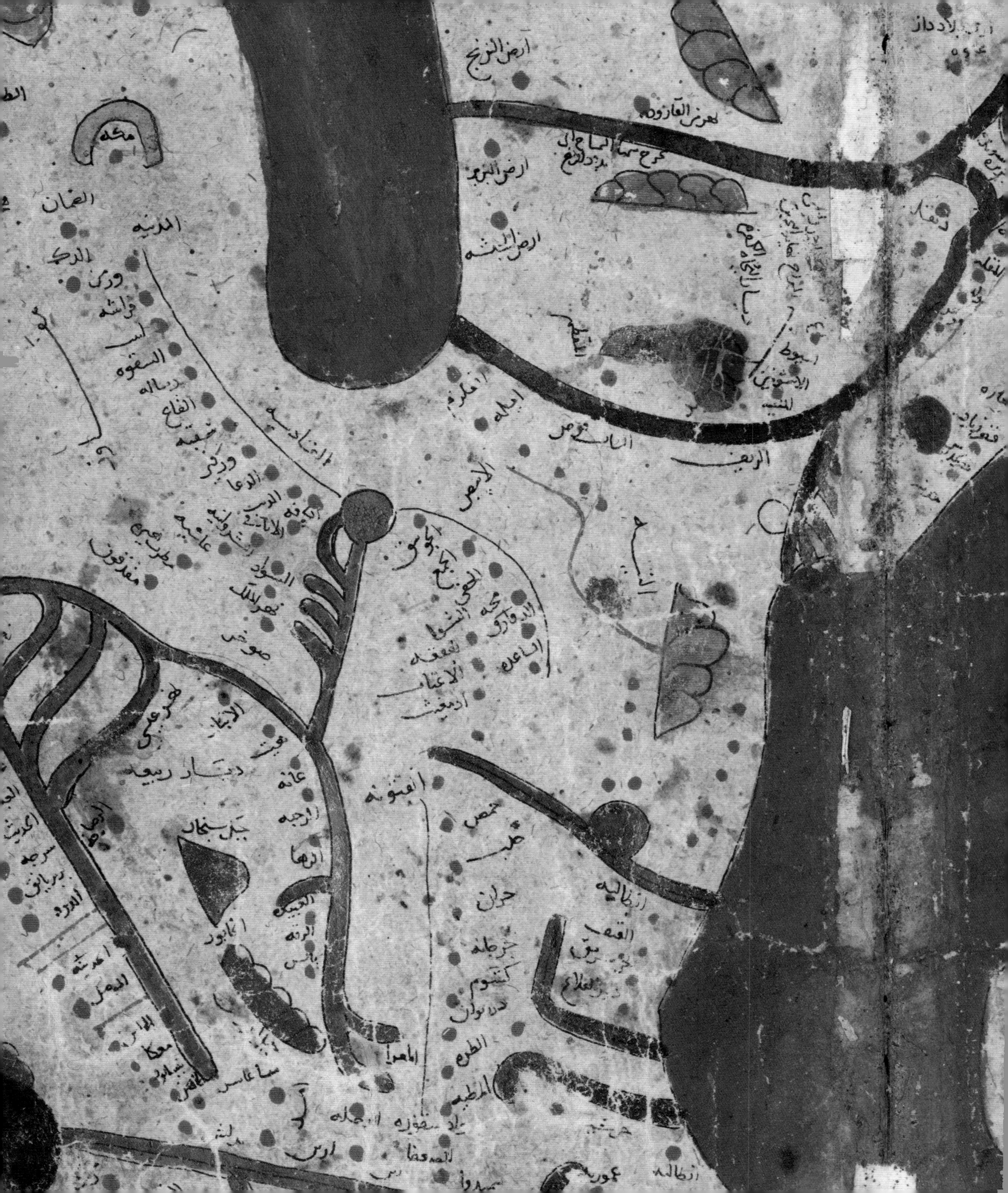

ارض الزنج
مكه
المدينه
ارض البربر
دنقله
اسيوط
الاشمونين
ديار ربيعه
الانبار
هيت
عانه
الرحبه
الرها
جبل سنجار
حمص
حلب
حران
انطاليه
عموريه

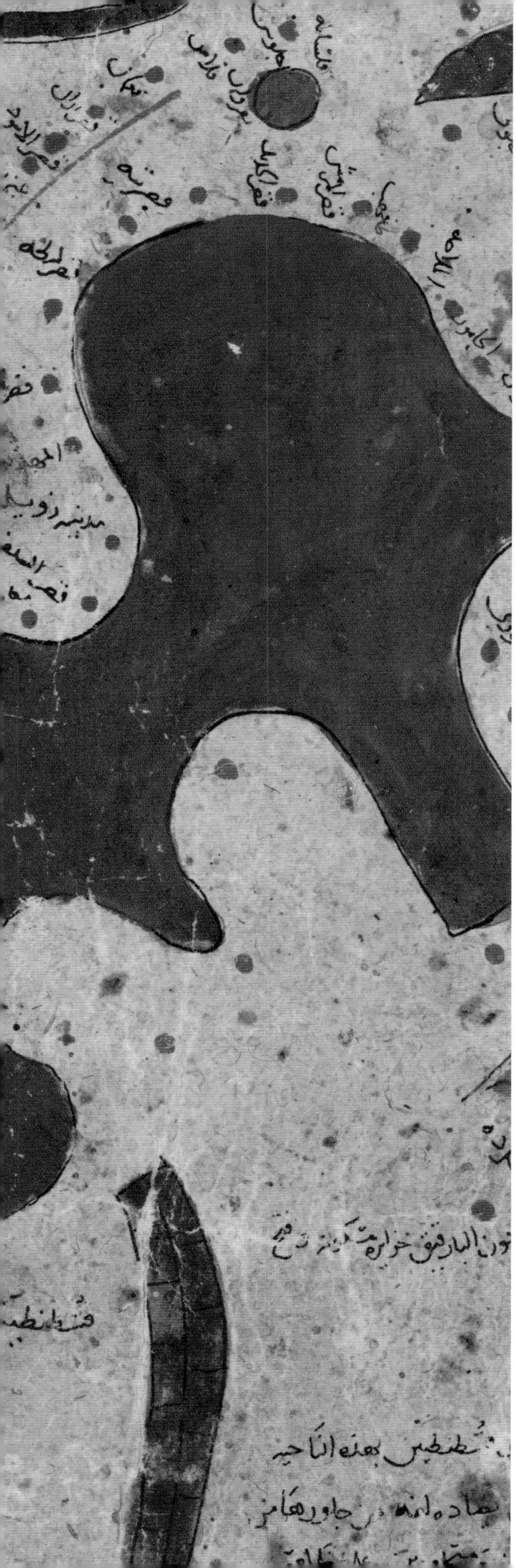

CHAPTER 1

A Mathematician's Map of the Nile

أوسيسه على البحر	سح ن	ز كه
طاس على بطحه	سط ه	نو ى
سوس على بطيحه	ع ك	نو مد
قسس	عا ه	نو مه

باماسوس على الجبل	فلاه	نح ى
باساموس جبل	قما ل	ما ك
مدينة ماجوج الداخلة	قعب ل	سح ه
طرمس على بطيحه	سطى	نه نه

الجبال التي خلف خط الاستوا

العدد	الاسماء	الحد الاول		الحد الاخر		ألوانها	جهات ذروتها
		الطول	العرض	الطول	العرض		
ا	جبل اورجلس	ح ل	ه ن	لح ل	د ه	اصفر	جنوب
ب	جبل بانقلوس	كح مه	ى له	لح م	و ل	لازورد	شمال
ج	جبل حسفادس	كح ه	ى كه	كح مه	ز ك	احمر	شمال
د	جبل انبيسقى	كد م	ه ك	ل ه	يا ى	لازورد	جنوب
ه	جبل بعضه في الاقليم الاول وبعضه خلف خط الاستوا	لد ى	ط كه	لد ل	ب مه	احمر	مغربي
و	جبل باريطون	لو ه	و ه	م ه	و ه	اصفر	جنوب
ز	جبل القمر ومنه مخرج نيل مصر	مو ل	ما ل	سا ن	يا ل	احمر	جنوب
ح	جبل اوله خلف الاستوا واخره في الاقليم الاول	مو ن	ط ن	ا ز	كح مه	حديدي	مغربي
ط	جبل الفيليا	ع ه	ه ه	[illegible]	د كه	حديدي	جنوب
ى	جبل على بيس اوله خلف الاستوا واخره في الاقليم الاول	عو ل	ب ه	ىعى ل	ا ك	اصفر	مغربي

The table of coordinates from al-Khwārazmī's *Book of the Image of the World* (*Kitāb Ṣūrat al-Arḍ*), copied 1037. Bibliothèque nationale et universitaire de Strasbourg, MS. 4247, fol. 10b.

EVERY MAP IS A unique mix of science and art, ideology and power. It is telling that the earliest Islamic map to have reached us was made by a mathematician – in fact, one of the greatest mathematicians of all times. Muḥammad ibn Mūsā al-Khwārazmī, who lived in Baghdad in the first half of the ninth century, is widely remembered today for his book *al-Jabr wa-al-Muqābalah* (*Reducing the Terms of an Equation by Addition and Subtraction*). In this book al-Khwārazmī explained, for the first time in the history of science, how to solve quadratic equations by using geometrical tools. The title of his book gave its name to the science we know today as 'algebra'. The title of the translation of the book into Latin, *Liber Alghoarismi* (*Book of al-Khwārazmī*), is the origin of the term 'algorithm'.

Al-Khwārazmī's interest in mathematics was triggered by the practical needs of the administration of a sprawling Abbasid empire, stretching from North Africa to the Indus Valley. Algebra was needed for calculating tax assessments and the division of inheritances, shares in water rights and engineering. Practical considerations also led to his interest in astronomy and in calendars. His astronomical tables, giving the location of hundreds of stars, were an important contribution to the systematization of timekeeping. He also left us an erudite study of the Jewish calendar, which fulfilled a similar purpose. Al-Khwārazmī believed in the ability of mathematical tools to control time, and also in the power of mathematical tools to represent geographical space.

The maps al-Khwārazmī drew are included in his *Book of the Image of the World* (*Kitāb Ṣūrat al-Arḍ*).[1] This is a book of geography, but a geography that takes no interest in narrative description of human societies. Rather, this is a mathematician's account of the world. For the most part, it consists of tables with lists of more than 4,000 place names and their locations, expressed in degrees of longitude and latitude. The places are arranged by categories, such as towns, mountains, seas, rivers and islands. For each town, the longitude in degrees and in minutes is given first, followed by degrees and minutes of latitude. Mountains, seas and rivers, which extend over large spaces, required more table columns. As seen in the folio pictured opposite, the name of each mountain is followed by columns for two sets of coordinates, one for the beginning of the mountain and one for its end. Another column indicates the colour of the mountain, and a final column records the direction of the mountain's peak.

This mathematical approach to the description of the world was the legacy of the great Greek astronomer and geographer Ptolemy, who worked in Alexandria in the second century CE. Al-Khwārazmī explicitly acknowledges his debt to the

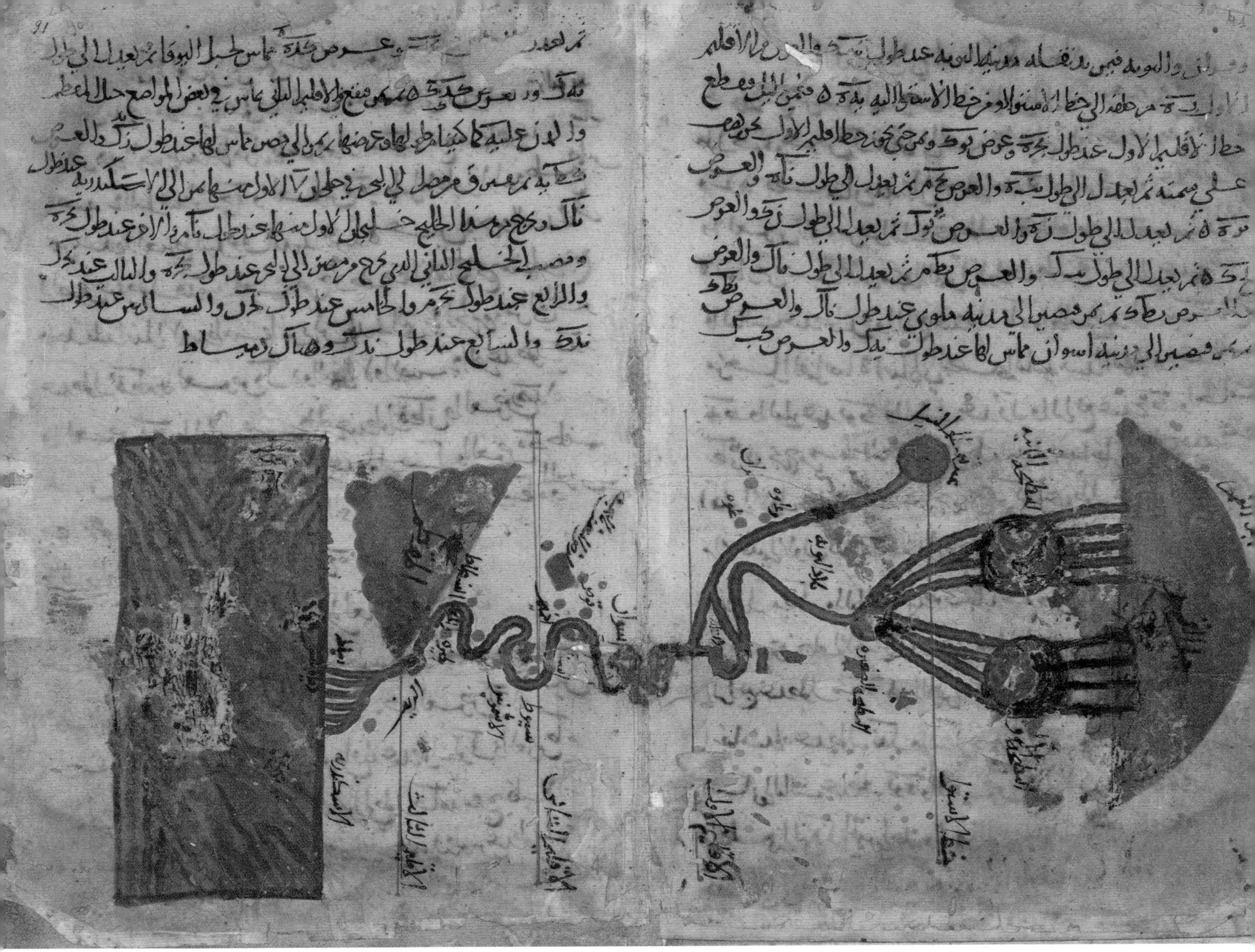

Map of the Nile from al-Khwārazmī's *Book of the Image of the World*, copied 1037. Bibliothèque nationale et universitaire de Strasbourg, MS. 4247, fols 30b–31a.

Greek master. Right on the first page, he states that his tables of coordinates are 'extracted' from Claudius Ptolemy's famous work, the *Geographia*. As we go through the text, it is clear that most of the place names in the tables are Arabic transliterations of Greek names, often doubly corrupted through the medium of a Syriac translation. But al-Khwārazmī owed Ptolemy much more than place names. It was the Greeks who established the spherical shape of the Earth and its approximate diameter. The Greeks also devised a method of dividing the northern hemisphere into seven horizontal bands, called climes. The first and southernmost clime was close to the equator, and its centre was in the city of

Meroë in Nubia, where the number of hours in the longest day was thirteen. The centre of the seventh and northernmost clime was the Dnieper River, where the longest day of the year lasted sixteen hours. This division of the northern hemisphere into seven climes was adopted by al-Khwārazmī in his own geography.

The *Book of the Image of the World* survives in one manuscript, copied in 1037, two centuries after al-Khwārazmī's death. This copy includes four maps, of which the most famous – and most elegant – is his map of the Nile (opposite). Since Arabic is written from right to left, we should begin to read the map from the right-hand side, where the page is dominated by the brown parachute-shaped mountain labelled 'Mountain of the Moon'. This is the source of the Nile, which al-Khwārazmī locates south of the equator. From the Mountain of the Moon, nine tributaries – five in the east and four in the west – flow northwards towards two large marshes. Another set of tributaries flow from these two larger marshes towards a smaller marsh, which then feeds into the main branch of the Nile. To the east, the map indicates another lake, labelled as 'Lake that feeds the Nile'. This is the source of an eastern tributary of the Nile. When this eastern tributary joins the Nile north of Aswan, a large island is formed at the confluence with the main branch of the Nile. Then the Nile meanders northwards towards the green square of the Mediterranean, passing the cities of Aswan, Asyut and Giza. The massive triangular mountain is the Muqattam Mountain. It towers over Fustat – the site of modern Cairo – in the same way as the mountain continues to dominate the landscape of Egypt's capital today. The river terminates in a schematic six-forked delta, flanked by the twin cities of Damietta and Alexandria.

The straight red lines that cut through the diagram are the indications of climes, harking back to the Greek system of dividing the inhabited parts of the northern hemisphere. On the right-hand side, the line closest to the Mountain of the Moon is the equator. The second line, passing just south of the large island formed by the confluence of the branches of the Nile, is the first clime. The second clime passes through Asyut, and the third clime through the Muqattam Mountain. It is easily noticeable that the distance between the clime lines narrows as we move northwards. This is not a mistake but reflects an acute mathematical understanding. Clime lines were defined by the length in hours of the longest day of the year, and the changes in the length of the longest day are more moderate as we approach the equator, becoming more extreme as we approach the North Pole. It means that the distance between the equator and the line of the first clime is about 16 degrees, while the distance between the

CAVRVS·CHORVS·VEL·IAPIX·SIVE·ARGESTES
CIRCTVS VEL·TRESIIAS
FAVONIVS·ZEPHIRVS
mare glaciale
EVROPA
LIBIA·INTERIOR
AFFRICA
Circulus equinoctialis
ETHIOPIA·INTERIOR
Barditi montes
Terra incognita secundum ptholomeum
Sinus hesperius
AFRICVS·VEL·LIBS
LIBONOTVS·EVROAVSTER

AQVILO VEL BOREAS
CECIAS APELIOTES
SVBSOLANVS
ASIA
Sinus magnus
Sinus gangeticus
MARE INDICVM
MARE INDICVM
PRASODVM
MARE
Tropicus cancri
Terra incognita secundum ptholomeum
Gradus longitudinis ab occidente in orientem
EVRNOTVS
WLTVRNVS EVRVS

first clime and the second is only 8 degrees. Remarkably, al-Khwārazmī's map attests to not only the division of the world into climes, but also the ability to convert clime boundaries into degrees of latitude. This is a map fully informed by mathematical concepts.

Al-Khwārazmī borrowed his conception of the origins of the Nile from the writings of Ptolemy, most of which had been translated into Arabic over the course of the preceding century. No map made by Ptolemy himself has survived, but a map based on his writings was made in Europe during the fifteenth century, as part of a renewed interest in the classical heritage (previous spread). The resemblance between the Nile in al-Khwārazmī's maps and the Nile in the maps based on Ptolemy is striking. The Mountain of the Moon and the twin marshes are easily identifiable with the same features as described in Ptolemy. The eastern lake, which is located on the equator, is undoubtedly a lake Ptolemy calls Lake Koloe (Κολόη), the source of an eastern tributary of the Nile. The confluence of the eastern tributary with the main branch creates the same large island. Ptolemy called it Meroë and, like him, al-Khwārazmī placed it at the centre of the first clime.[2]

Yet al-Khwārazmī was not a slavish copyist. To the twin marshes that Ptolemy thought were the sources of the Nile, he added a third, smaller lake, which in his map effectively becomes the source of the main branch of the Nile. Al-Khwārazmī also completely updated the names of the towns along the course of the Nile. For example, he correctly located the trading town of Qūṣ north of Aswan. In the Islamic period, Qūṣ became an important trading hub on a desert route that linked the Nile with the Red Sea. He also added the Muqattam Mountain and the town of Fustat, the garrison town established by the conquering Arabs in the seventh century. During al-Khwārazmī's days, Fustat was the capital of Egypt, and it remained so until the foundation of Cairo in 971.

What al-Khwārazmī didn't do here, or anywhere else in his writings, was to create a plotted map. There are clime lines, which indicate approximate latitudes. But there is no scale, no hint of degrees. Outside the map, the adjoining text gives numerical values of the longitude and latitude of most places listed on the map – the numbers are given in a system that uses the Arabic alphabet, but they are numbers all the same. The map below, however, only illustrates the mathematical data, without formally plotting it on a grid. Nowadays we would expect al-Khwārazmī to use the numerical values to plot the course of the Nile and the towns along its banks. This, we tend to think, would turn the diagram of the Nile into an 'accurate' map based on the mathematical tools that he seems to

(previous spread) World map from a translation of Ptolemy's *Geographia*, 1486. Bodleian Library, University of Oxford, Arch B b.19, fols 131v–132r.

prize so much in his tables of coordinates, which fill the rest of the book. But al-Khwārazmī didn't leave us such a map. And, while that has proved disappointing to many modern observers, we should not view this absence as a shortcoming.

There seems little doubt that al-Khwārazmī, the progenitor of the science of algebra, had the mathematical ability to make a plotted map. A century after al-Khwārazmī, another scientist called Suhrāb offered illustrated instructions for the construction of a world map on a rectangular grid. We know of Suhrāb only from one treatise he wrote, entitled the *Book of the Marvels of the Seven Climes* (*Kitāb ʿAjāʾib al-Aqālīm al-Sabʿah*). This book, like al-Khwārazmī's *Book of the Image of the World*, consists mostly of tables of coordinate values – degrees of longitude and latitude – for cities, mountains, seas, rivers and islands. The data are very similar to those provided by al-Khwārazmī and, as in al-Khwārazmī's work, the tables are arranged according to the seven climes.

Most importantly, Suhrāb explicitly linked the coordinate values with the construction of a rectangular map of the inhabited parts of the northern hemisphere. He stated that a world map was to be drawn on a large rectangle, with degrees of latitude and longitude marked along the edges. On its northern and southern sides it should show 180 degrees of longitude, and on its western and eastern sides the scales should run from 20 degrees south to 90 degrees north; then clime lines should be drawn according to latitude values. After constructing the frame of the world map, individual towns, mountains and rivers were to be plotted with a pair of weighted strings. The location of these place names, in degrees of longitude and latitude, was taken from the coordinate values in the main body of Suhrāb's treatise. A diagram that accompanies this explanation (p. 20) shows how the edges of the rectangle should be marked in divisions of 10 degrees of longitude and latitude, and how the lines of the climes and the equator should be indicated.

Suhrāb tells us how to create a mathematically plotted map of the world using a simple rectangular projection. This most basic form of projection of the spherical surface of the Earth onto a flat surface is quite problematic, as it results in substantial distortions when we move away from the equator and towards the northern regions of Europe and Asia, as well as on the eastern and western edges of the map. This type of simple projection was known to Ptolemy, who criticized it severely. The great Muslim scientist al-Bīrūnī, who lived in the eleventh century, also dismissed this projection as too crude. Al-Bīrūnī, by the way, devoted an entire treatise to projections, where he offers much more sophisticated

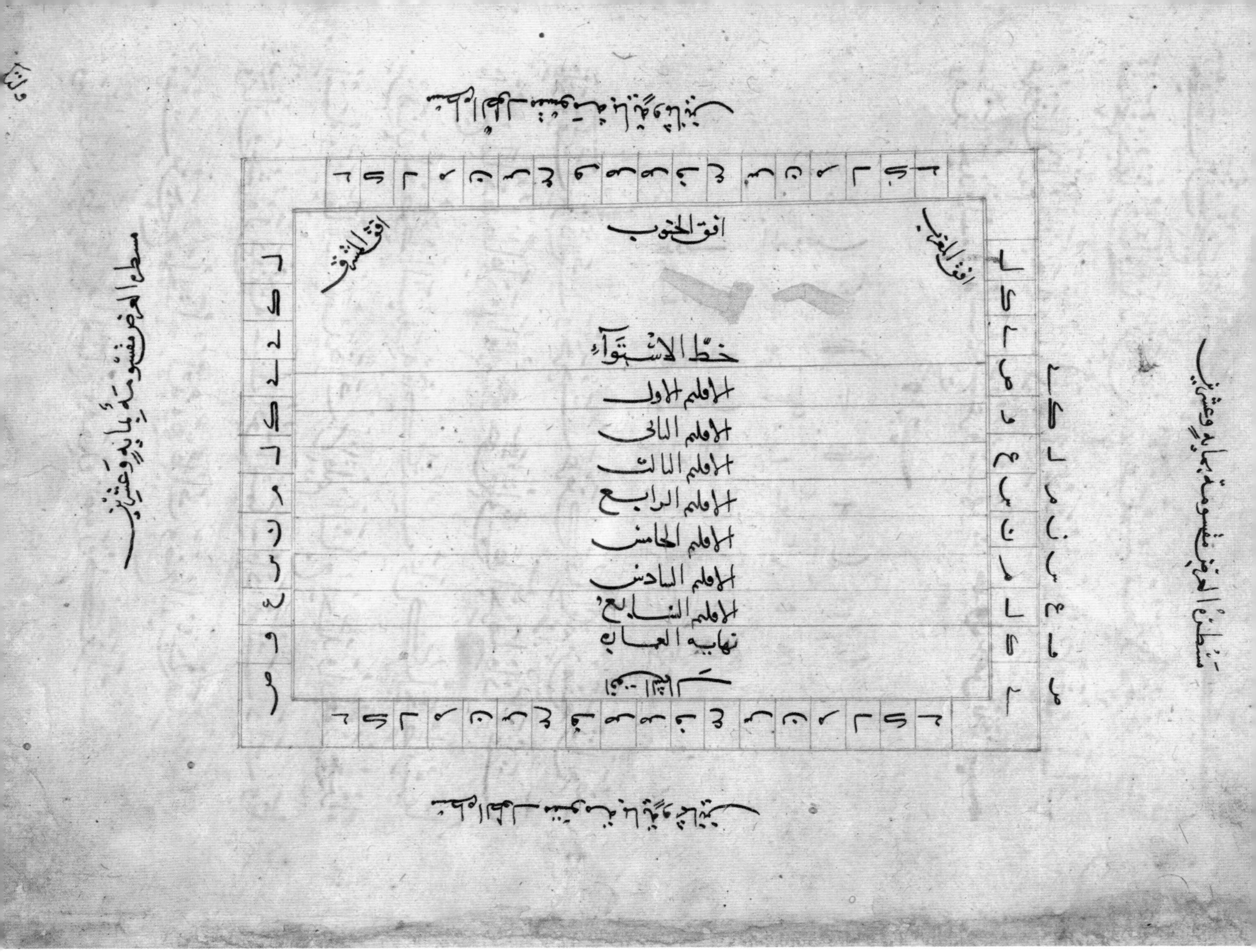

alternatives that take into account the curvature of the globular Earth. Yet, for all their instructions concerning projections, neither Suhrāb nor even al-Bīrūnī left us with a mathematically plotted map.

Some modern scholars argue that al-Khwārazmī's *Book of the Image of the World* must have been accompanied by a world map that has now been lost. Moreover, they often argue that the world map attached to his book was created at the order of the Abbasid caliph al-Maʾmūn, a champion of Greek sciences and rational thought in the first half of the ninth century. Al-Maʾmūn had a keen interest in geography, and famously commissioned a group of astronomers and surveyors to

Suhrāb's diagram for constructing a world map, in his *Book of the Marvels of the Seven Climes* (*Kitāb ʿAjāʾib al-Aqālīm al-Sabʿah*), copied 1309. © The British Library Board, Add. MS. 23379, fol. 4b.

measure the extent of 1 degree of the circumference of the Earth. They did this by comparing the altitude of the sun at midday in two known locations in the Plain of Sinjar, in present-day Iraq. The measurement they obtained, 56⅔ miles for 1 degree of latitude, results in the remarkably accurate figure of 20,400 miles for the circumference of the Earth as a whole.

Al-Maʾmūn also commissioned a map of the world, beautifully described by the tenth-century polymath al-Masʿūdī. Al-Masʿūdī tells us that he has seen three world maps, all showing the seven climes of the inhabited parts of the northern hemisphere. Two of these world maps were made by Greek scientists before Islam, and the third was the one made for al-Maʾmūn. The latter, he claims, was far better than the maps of late antiquity:

> I have seen these climes represented without labels and in different colors. The best that I have seen is in the *Geographia* of Marinus and in the commentary on the *Geographia* of the divisions of the Earth, and in the map of al-Maʾmūn. That is the map made for al-Maʾmūn, which was constructed by a group of contemporary scholars, and in which the world is represented with its spheres, stars, lands and seas, the inhabited and uninhabited regions, settlements of peoples, cities and so forth. This [map of al-Maʾmūn] was better than anything that preceded it, either the *Geographia* of Ptolemy, the *Geographia* of Marinus or any other.[3]

Al-Masʿūdī saw at least three world maps that indicated clime boundaries: two were versions of the works of Ptolemy and Marinus (another Greek scholar who worked in Tyre in the first century CE) and the third was drawn for the caliph al-Maʾmūn. Al-Masʿūdī states that the one made for al-Maʾmūn compared favourably with the previous ones, as it contained rich details on the 'spheres, stars, lands and seas'. Al-Maʾmūn's map must have been an impressive sight, perhaps even for public display, and its loss means that a key moment is missing from the history of Islamic maps. But there is nothing in this account that proves that the map made for al-Maʾmūn was mathematically plotted, nor is there any mention of al-Khwārazmī. We can only say that, like al-Khwārazmī's map of the Nile, it indicated the boundaries of the climes and that it was informed by astronomical measurements. The claims that al-Khwārazmī's tables and maps were the blueprint for al-Maʾmūn's map of the world are merely tangential. It is true that al-Khwārazmī could produce a plotted map of the world, but that doesn't mean that he actually did so.

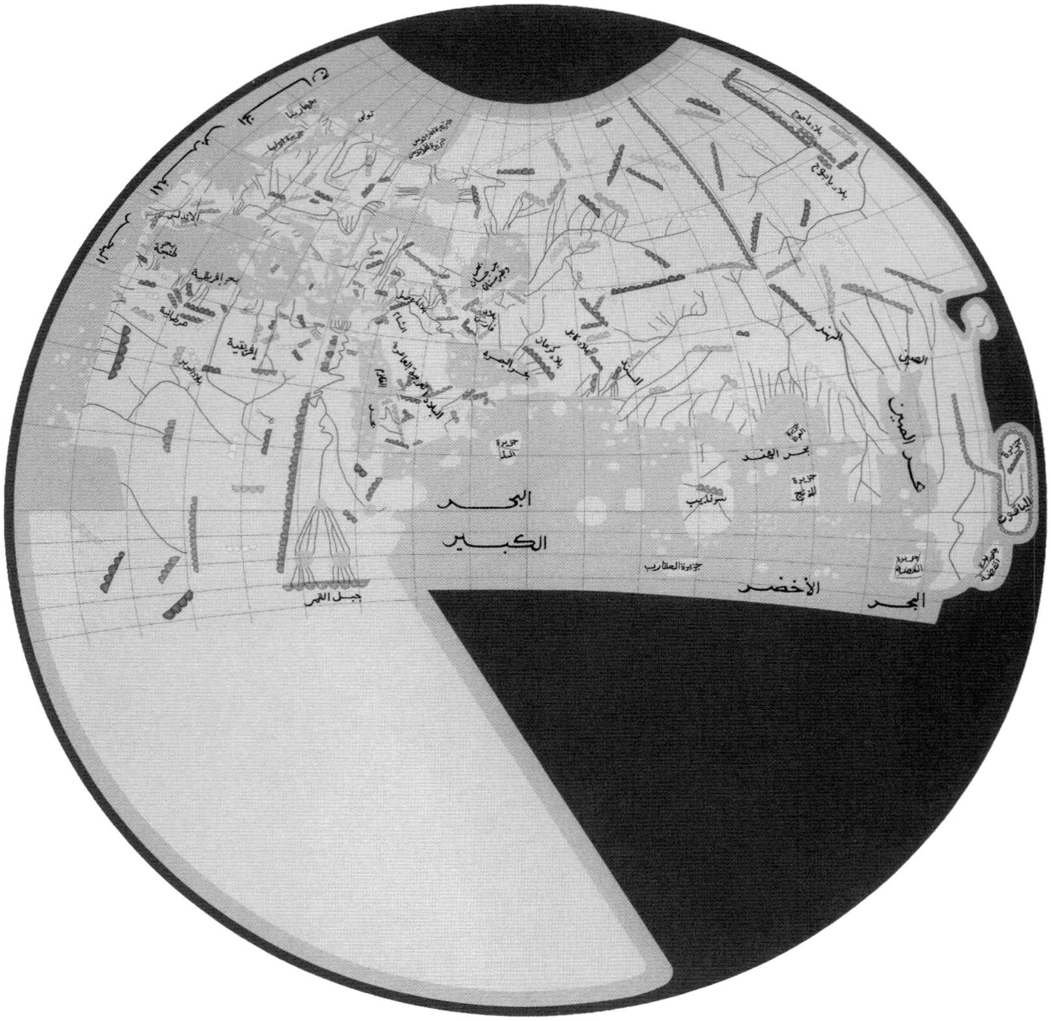

The Grammar of Islamic Map-Making

So what did al-Khwārazmī try to achieve with his map of the Nile? The illustration is clearly informed by the latitude and longitude data given in the text. In recent decades, there have been several attempts to construct a world map based on the coordinates al-Khwārazmī provided in his *Book of the Image of the World*. The map created by Fuat Sezgin, which uses modern technologies of projection, is the most elaborate attempt (above).[4] It is important to keep in mind that this is a reconstruction, employing modern techniques and appealing to modern expectations, and not how a map made by al-Khwārazmī would have looked. Nonetheless, the resemblance between the Nile map actually made by al-Khwārazmī and the map reconstructed by Sezgin from the coordinate values in al-Khwārazmī's tables is striking, especially the system of lakes at the origins of the Nile and the position of the clime lines.

Reconstruction of a world map, using modern projection, on the basis of coordinates provided by al-Khwārazmī in his *Book of the Image of the World*, prepared by Fuat Sezgin, 2000. From Fuat Sezgin, *Mathematical Geography and Cartography in Islam and their Continuation in the Occident*, 3 vols, Institute for the History of Arabic-Islamic Science, Frankfurt am Main, 2000–7, vol. III, plate 1b. Courtesy of Institut für Geschichte der Arabisch-Islamischen Wissenschaften.

The differences between the medieval Nile map and the modern reconstruction are telling. Al-Khwārazmī, or the eleventh-century scribe who copied the work, opted for legibility over accuracy, for artistic effect over mathematical precision. One important decision was to condense the upper regions of the Nile relative to the more populated northern areas. This allowed him to introduce, within the confines of the limited space of the page, many more labels representing the dense network of Egyptian cities. The sparsely populated and relatively unknown areas of Nubia and the equatorial Nile are mostly blank. Since not many labels are required, the size of these remote regions is fittingly reduced. As is typical in medieval maps, the *terra incognita* has been made to look smaller than the lands that are well known. Moreover, the Egyptian lands are now emphasized by the oversized Muqattam Mountain, so familiar to the Muslim elites in charge of the Egyptian capital of Fustat. Al-Khwārazmī's map of the Nile was informed by the mathematical tools of longitude and latitude, but consciously eschewed employing them, privileging the effect the map would have on the viewer over the precise replication of the data.

Al-Khwārazmī was not making plotted maps. Instead, he was trying to create a grammar of Islamic map-making in the same way as he created the fundamental language of algebra. This is most apparent in another map, or diagram, included in his *Book of the Image of the World* (p. 24). The text above the illustration states that this is a map of the Sea of Darkness, which is the name al-Khwārazmī and Muslim geographers after him gave to the ocean that encircled the entirety of the land masses known to them. But the illustration doesn't actually represent any specific geographical space, and none of the labels name an individual land or town.

Rather, this is a sketch map in which al-Khwārazmī merely illustrated the manner in which a map-maker should represent coastlines. The labels here only describe the types of bays, gulfs and headlands that are being depicted pictorially. For example, all the headlands in this image, some circular and some pointed, are labelled *quwārah* (قوارة). The word literally means the neck of a bottle, but is here used in a technical way, as part of a new nomenclature of cartography. This, al-Khwārazmī tells us, is how a *quwārah* should look like. All three pointed triangular-like gulfs in this illustration are labelled *shābūrah* (شابورة), and are distinguished from the wider circular bays in the four corners of the diagram, all labelled *ṭaylasān* (طيلسان). In common Arabic usage the word *ṭaylasān* refers to the broad shawl worn by religious scholars but again it is employed as a technical term for the use of map-makers and map copyists.

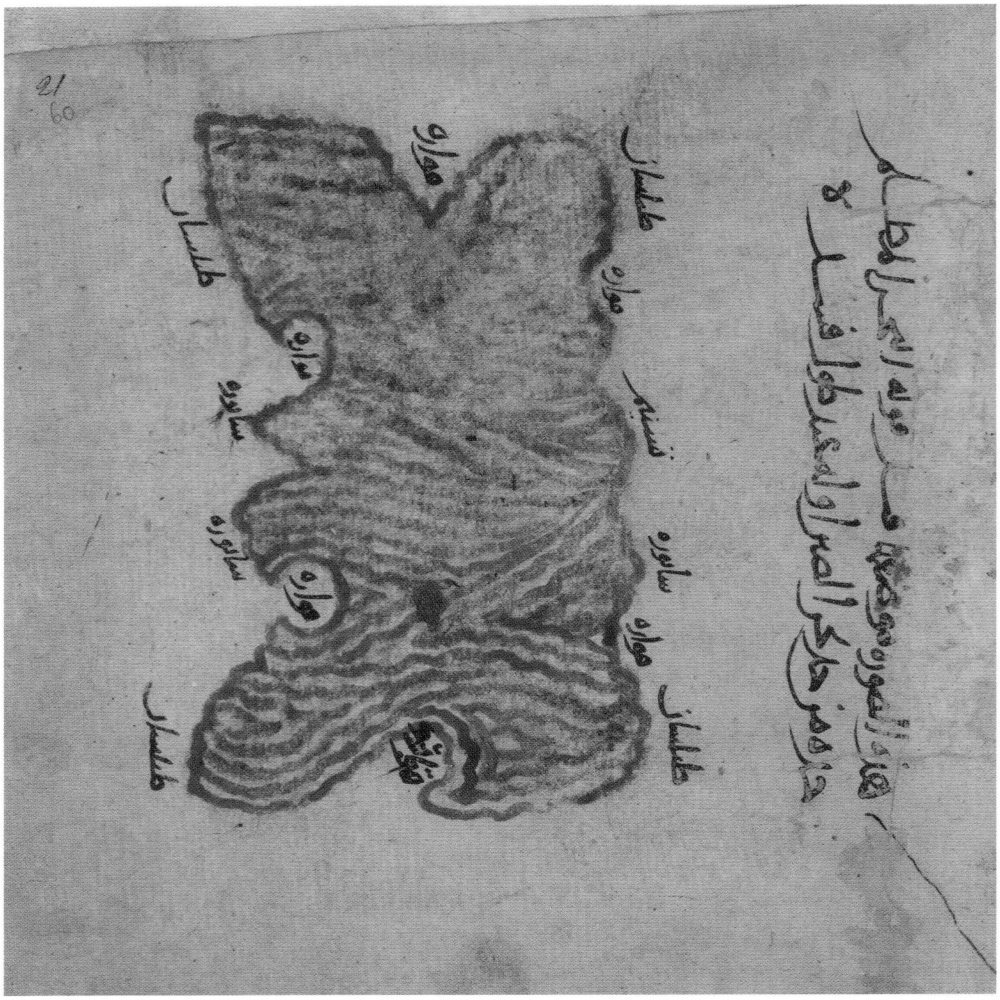

(left) Sea of Darkness from al-Khwārazmī's *Book of the Image of the World*, copied 1037. Bibliothèque Nationale et Universitaire de Strasbourg, MS. 4247, fol. 21a.

(opposite) Detail from the map of the Nile from al-Khwārazmī's *Book of the Image of the World*, copied 1037 (p. 14). Bibliothèque Nationale et Universitaire de Strasbourg, MS. 4247, fols 30b–31a.

By linking text and image in this way, al-Khwārazmī gave generations of Muslim cartographers after him a rudimentary symbolic language for producing maps. The technical terms *quwārah*, *shābūrah* and *ṭaylasān* are used throughout his *Book of the Image of the World* whenever he describes the shapes of the seas. The sketch here illustrates what these coastal features look like and how they should be drawn. These pointed gulfs and concave headlands would, he hoped, become the building blocks of Islamic map-making. We must appreciate the ingenuity of al-Khwārazmī facing the challenge of copying maps in a manuscript culture. Before the introduction of print, copyists were unable to faithfully reproduce every detail of a map they encountered. Rather than asking copyists to perform an impossible task, al-Khwārazmī gave them the cartographic language to create new maps that would convey the same visual meaning as the maps they were trying to copy.

Recognizing al-Khwārazmī's ingenuity is not to deny his debt to earlier map-makers. Al-Khwārazmī was utilizing pre-Islamic materials and concepts, particularly those that came through Greek works translated into Syriac in

the fourth, fifth and sixth centuries. These Syriac translations, it seems, included maps, and it is very likely that al-Khwārazmī compiled the tables of coordinates in his *Book of the Image of the World* by looking at one such map. In fact, he tells us several times that he is unsure about the name of a certain locality because the writing on the 'map' (*ṣūrah*) is not sufficiently legible.[5] He thought that the map he was utilizing was constructed by Ptolemy himself, although in fact he was probably using an adaptation made in late antiquity. In any case, this is probably what he means when he says that his tables have been 'extracted' from the *Geographia* of Ptolemy – he had looked at the late-antique world map and extrapolated the coordinate values of different topographical features. Mathematician that he was, al-Khwārazmī transformed a world map into tabular form, in the same way as a modern scientist might break down the analogue complexities of nature into Excel files.

Mapping the Nile – Again

Al-Khwārazmī's work survives in only one copy, made two centuries after his death. This is partly a result of the lukewarm reception it was given by the majority of later Muslim geographers and map-makers. As we shall see in Chapter 2, his mathematical approach to the depiction of geographical space was rejected for the most part. But al-Khwārazmī did have his imitators, and the best example comes from a recently discovered eleventh-century treatise on cosmography and geography known as the *Book of Curiosities*, which is now housed in the Bodleian Library. This treatise contains a map of the Nile that is, beyond doubt, a variant of the map of the Nile by al-Khwārazmī (see p. 27 and above). Even though the lower left part has been lost because of damage to the page, the missing parts can be reconstructed with the help of the text on the side panel, which describes the course of the Nile as it appears on the map.[6]

The Nile maps in the *Book of Curiosities* and in al-Khwārazmī's treatise are so similar that it takes time to spot the differences. The dominant features of both maps are the parachute-shaped Mountain of the Moon and the system of three lakes at the sources of the Nile, connected by parallel sets of outlets. The clime lines are another evident similarity. In both maps the equator passes between the twin lakes and the third, northern, lake, and the first clime is noticeably wider than the second clime. In the Nile map of the *Book of Curiosities* the upper half of a blue lake (blue indicating sweet water) on the left is undoubtedly the same as the eastern lake of al-Khwārazmī's map. The lost part of the map included an eastern tributary, the equivalent of the eastern tributary indicated on al-Khwārazmī's map. The side panel explains that '[The Nile] is joined by a river coming from the land of Zanj [East Africa], from a lake that is called the "Flask" and is also known as Lake Qanbalū.'

The most innovative element in this second Nile map, and the one that had the most profound influence on the development of later Islamic cartography, is the depiction of a western tributary of the Nile that flows from 'white sand dunes' in West Africa. On the map the sand dunes are indicated by a red mountain located on the equator to the west of the Nile, on the right-hand side. The label near the mountain reads 'White sand dunes from which a river flows to the Nile'. In the the side panel, we read about a tributary issuing from a spring that emerges beneath white sand dunes along the Atlantic shores of North Africa. This conception of a western tributary of the Nile emerging in North Africa is found in Greco-Roman geographical literature. Pliny the Elder, for example, related that the Nile rises in Lower Mauritania, not far from the Western Ocean. His account was later reclothed in Islamic garb through stories about Arab conquerors of North Africa who encountered a huge sand dune in the westernmost parts of North Africa, and who realized that the Nile issues forth from these massive sand dunes towards Egypt.

The map of the Nile in the *Book of Curiosities* also acknowledges mathematical tools in a way that goes beyond al-Khwārazmī's maps. Very unusually, some of the labels on the Nile map in the *Book of Curiosities* include data on longitude and latitude. For example, the rivers that flow from the Mountain of the Moon are said to be located between 46 degrees and 59 degrees longitude. Both the diameters of the lakes at the source of the Nile and the distances between the tributaries of the Nile are given in units of 'celestial' degrees, which are then converted into miles. For instance, the diameter of the eastern marsh is given as 'five degrees, equivalent to 284 miles' – which is what we would expect if we take each degree of

(opposite) Extant portion of the map of the Nile in the *Book of Curiosities*, an 11th-century Egyptian treatise on cosmography and geography, copied *c.*1200. South is at the top. Bodleian Library, University of Oxford, MS. Arab c. 90, fol. 42a.

صورة النيل

ومخرجه من عشرة انهار في جبل القمر منها في مشرقه خمسة وفي مغربه خمسة ثم ينصب الى البحيرتين ومن البحيرتين الى بطيحة كبيرة على خط الاستوا ثم يجري الى الثانية الثمان وباتيه من نحو بلاد البجة من جري بنيل الها

القارورة وتعرف ببحيرة قيلوا وباتيه بنهر اخر من نحو بلاد المغرب من ما يحبس تحت الكثيب الابيض عند ناحية البحر المحيط ويصب فيها انهار كثيرة ثم يمتد جريه عند نقصان كل نهر على وجه الارض حتى انه لو قال قايل

انما تمر في زيادته بمبايها كان ذلك ومطالع القرطان والتاسع المريخ ومعرفة زيادته ان تنظروا الى طالع السنة الى المريخ فان كان في مبرحه الاكبر كانت الزيادة موثرة وان كان في مبرحه الاوسط تمته

زيادته وان كان بطيئا في مبرحه كان ناقص الجري فاعلم ذلك

مكه
المدينه
الطايف

the circumference of the Earth to be 56⅔ miles, following the calculations made by the scholars commissioned by al-Maʾmūn. This second Nile map is not plotted, but it is again heavily informed by a mathematical conception of space. It locates individual features in relation to an invisible grid of degrees of latitude and longitude, and, like scales on modern maps, it translates the notional degrees into actual distances.

In the *Book of Curiosities* we also find the earliest surviving world map to carry a scale – not just in Islam but in any culture (previous spread). This spectacularly enigmatic rectangular map of the inhabited world is so strange that most modern viewers would not immediately recognize it as a world map at all. Even after we realize that south is at the top, the relative location and size of the continents are unfamiliar. Europe, at the bottom right, is an enormously disproportionate island, mostly taken up by an oversized Iberian Peninsula. Asia, in contrast, is tightly squeezed at the left side of the map, while the Arabian Peninsula is disproportionately prominent. Mecca is indicated by a distinctive yellow horseshoe symbol, the only locality not indicated by a simple red dot. Most of Africa is off the map.

Closer observation reveals the distinctive parachute-like Mountain of the Moon, a source of the Nile, at the top of the map and the southernmost point shown. This visual element is taken directly from al-Khwārazmī's iconography, and is replicated in the map of the Nile of the *Book of Curiosities*, which we have just discussed. There is no doubt that all three maps are related. But while al-Khwārazmī's map is focused on the course of the Nile, here we zoom out to see the world as conceived by a map-maker employing some of the tools of mathematical geography.

What makes the world map of the *Book of Curiosities* unique, and what got historians of cartography all excited when this manuscript came to scholarly attention at the beginning of the twenty-first century, is the scale at the top of the map (opposite). This scale, which has been carefully executed, has been drawn as part of the outline of the map. Examination by infrared and ultraviolet lamps revealed that some of it lies under the green paint of the ocean and the brown Mountain of the Moon.[7] The cells on the right-hand folio are numbered with *abjad* letter-numerals. These letters of the Arabic alphabet give numerical values, increasing in from 5 degrees in the top right to 135 degrees, the last visible number before the scale is overpainted with the Mountain of the Moon.

To be sure, the scale is significantly corrupted. The most obvious error is that the scale on the map is divided up so as to reach 360 degrees, as if it represents the

(previous spread) A rectangular world map from the 11th-century *Book of Curiosities*, the earliest surviving world map to feature a scale, copied *c.*1200. South is at the top. Bodleian Library, University of Oxford, MS. Arab c. 90, fols 23b–24a.

Detail of the scale bar on the rectangular world map from the *Book of Curiosities*. Bodleian Library, University of Oxford, MS. Arab c. 90, fols 23b–24a.

entire Earth. But the map only shows the inhabited land masses of the Earth, from the Atlantic to China, which are assumed to occupy half of the Earth's surface. Therefore, the scale should have shown only 180 degrees across the map. This mistake renders the scale as it appears here meaningless, almost decorative. The map-maker, or at least the copyist who produced the copy that we have, did not have a good grasp of the way this longitude scale was to be used. It is also evident that the scale has no bearing on the position of any of the localities. Even after the numbering of the scale is adjusted to the correct 180 degrees, the location of the most prominent Islamic cities, such as Córdoba, al-Qayrawan and Mecca, is not at all aligned with the longitude values found on the scale. There is also no consistent alignment with the expected latitude for the same cities. The scale at the top notwithstanding, this world map was not mathematically plotted either.

But, then again, the conception of the map is informed by mathematical tools and by the notions of longitude and latitude. The anonymous author of the *Book of Curiosities* explains that he intends his world map to show the inhabited part of the world, from the equator to the 'farthest limit of the inhabited world, which is at 66 degrees [of latitude]', as is related in the *Geographia* of Ptolemy. He tells us that his map aims to show only the inhabited world, the inhabited quarter of the surface of the Earth. As we go back and look at this world map, we see that the scale most probably runs along the equator, right at the top, while the uninhabited parts of Africa beyond the equator are left out. At the bottom the map shows northern Europe, but does not go as far as the North Pole. Instead, it is bounded by a line of latitude that is somewhere north of al-Khwārazmī's seventh clime.[8] At the bottom left we find a large wall, and the label explains that this wall prevents the people

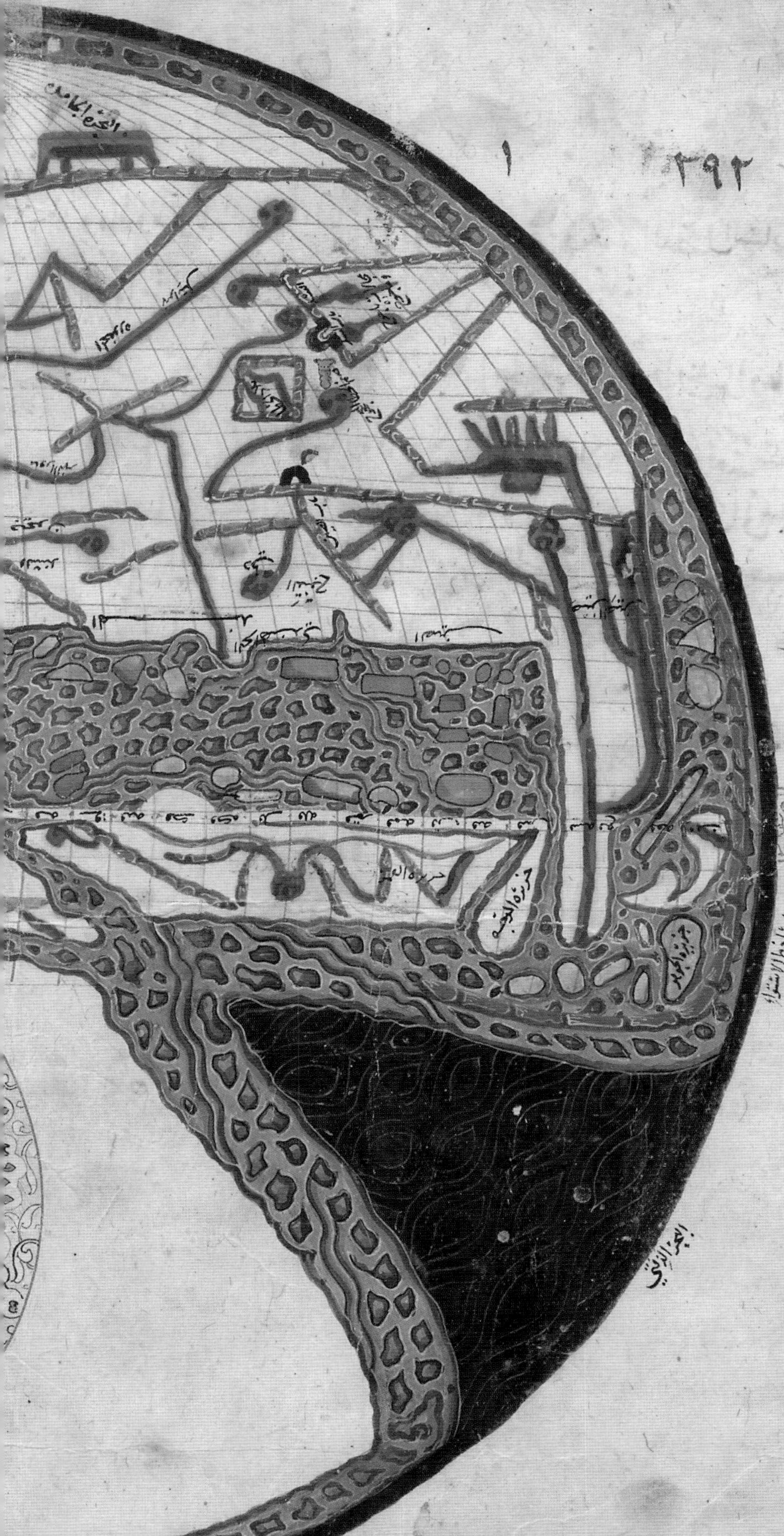

of Gog and Magog from encroaching on human civilization. This wall marks the end of the inhabited world. Al-Khwārazmī located this wall of Gog and Magog at 63 degrees north.

The unusual rectangular shape of this world map is itself a product of the way in which its boundaries are framed by degrees of latitude. Islamic map-makers, like most of their European contemporaries, generally chose to present the Earth as a circle or disc, and we shall see many examples of circular world maps in the following chapters. The *Book of Curiosities*' world map is unusual in that it represents only the inhabited parts of the world, which are defined by degrees of latitude – from the equator to 66 degrees north. Once the North Pole and any area south of the equator are excluded, a rectangular shape is simpler to execute. We should recall here the illustrative diagram drawn by Suhrāb and his simple projection. A rectangle fitted better with the mathematical conception of this map.

(previous spread) A world map from an encyclopedia written by the 14th-century Egyptian author Ibn Faḍlallāh al-ʿUmarī, *c.*1340. Topkapi Palace Museum, Istanbul, MS. A 2797, fols 292v–293r.

A Plotted World Map at Last?

We conclude the journey that began with al-Khwārazmī's coordinate tables with a fourteenth-century world map that is the epitome of mathematical map-making in medieval Islam (previous spread). This remarkably complex circular world map, included in an encyclopedia written by the fourteenth-century Egyptian author Ibn Faḍlallāh al-ʿUmarī, has a scale running along the equator, as in the rectangular world map in the *Book of Curiosities*. The author attributes the map to the 'author of the *Geographia*'. But, since some of the few labelled cities (such as Delhi) indicated on the northern coasts of the Indian Ocean were not established till the thirteenth century, this could not have been a direct copy of a map made by either Ptolemy or al-Khwārazmī.[9] In this map, the degrees on the scale placed on the equator are correctly marked, and there are lines marking divisions of longitude that run from the equator towards the North Pole. There is no similar scale for degrees of latitude, but the legend outside the map enumerates the seven climes north of the equator. There is nearly a complete grid, and the shapes of the continents are more readily recognizable, thanks partly to more sophisticated projection. Finally, we have arrived at a map that may have been mathematically plotted, at least in part.

Why did the mathematical approach to map-making meet with such limited and belated success among cartographers working in the medieval Muslim world? Al-Khwārazmī provided tables with longitude and latitude for thousands of locations, and many astronomers and geographers after him added and improved on his tables. Suhrāb showed us that the principle of creating a world map by plotting latitude and longitude values on a grid was easily understood. The mathematical capability to construct plotted maps was evidently already there by the ninth century. And yet, save perhaps for the fourteenth-century map of al-ʿUmarī, no plotted map has survived. This is not a coincidence of the chance preservation of manuscripts – hundreds of other maps do survive – nor is it due to some failure of the medieval imagination. It was due to a conscious rejection on the basis of practical and ideological reasons.

The truth is that, in the medieval context, plotted maps were terribly impractical. The accuracy of maps depends on the accuracy of the information that is placed on them. Although the tables of latitude and longitude values provided by al-Khwārazmī and others are impressive in their ambition, they are simply not very accurate. The tools to measure latitude, such as by measuring the location of the sun at midday or the length in hours of the longest day, gave only approximate values. Measurements for longitude were even cruder. For all their scientific allure, the tables of coordinates were very rough estimates, useful for timekeeping or for finding the prayer direction towards Mecca. The margin of error was so large that there was no point whatsoever in plotting them on sectional maps of provinces. The only scale on which they could have made sense was on world maps.

But then comes the problem of projection, which is much more acute in world maps than it is in sectional maps. In maps of smaller areas, such as a map of Egypt or of England, transferring the curvature of the Earth to the flat surface of the paper does not create significant distortions. But, in trying to represent the Earth or even just the northern hemisphere, the distortions are so considerable as to require complex mathematical solutions. A simple rectangular projection as suggested by Suhrāb would have meant that the northern, western and eastern edges of the map were stretched and made to look much larger than they actually were.

Even if such problems of projection could have been solved, what purpose would such world maps serve? We do know that the eleventh-century scholar al-Bīrūnī proposed several projections that would have resulted in more accurate world maps. But his own world map is also merely a sketch. World maps are particularly useful for travelling in the open oceans, and it is no coincidence that the first world maps to represent distances and continents with some degree of accuracy accompanied the European age of exploration. Muslim empires were mostly land based, and when Muslim merchants travelled in the

Mediterranean or in the Indian Ocean they stayed close to the coast or relied on the predictable monsoons.

Muslim map-makers, starting with al-Khwārazmī himself, privileged legibility and effect over precision. The mathematical tools that originated with Ptolemy were known, but they were consciously and explicitly put aside. We know this from the direct testimony of one map-maker, the anonymous eleventh-century Egyptian author of the *Book of Curiosities*. After providing the rectangular map of the world discussed above, with a scale at the top as a gesture to degrees of longitude, the author makes a clean break from that tradition. The maps that follow, he says explicitly, are not meant to be accurate representations. He gives two reasons for his decision to eschew accurate depiction of coastlines. First, he argues that land can turn to sea and sea to land, citing several examples from the not-so-remote past, when earthquakes and gigantic waves permanently moved the coastlines. Over time, he says, coastlines shift through the forces of nature, and maps that are too precise become out of date.

Second, and this is where the limitations of mathematical map-making are most clearly exposed, plotting coastlines by coordinates is not in tune with the needs of those using the maps. Referring to the terms used by al-Khwārazmī to describe the building blocks of mathematical map-making, he states that maps drawn in the manner described by Ptolemy in his *Geographia*, have coastline-forming 'curves in the coast [*ʿaṭfāt*], pointed gulfs [*shābūrāt*], square [*murabbaʿāt*] and concave headlands [*taqwīrāt*]'. He continues:

> This shape of the coast exists in reality, but, even if drawn by the most sensitive instrument, the map-maker [*muhandis*] would not be able to position [literally, 'to build'] a city in its correct location amidst the curves in the coast or pointed gulfs. This is because of the limits of the space [on the page], as against the vast area in the real world. That is why we have drawn this map in this way, so that everyone will be able to figure out [the name of] any city.[10]

It is perhaps possible, our anonymous map-maker says, to construct plotted maps using delicate instruments and precise measurements, in the manner recommended by Ptolemy. But this would defeat the purpose of maps, because the irregularity of the coastline would not leave room for labels to be placed. Here we have it from the horse's mouth – a medieval Muslim map-maker opts to eschew precision in favour of legibility, and does so explicitly out of considerations of functionality. There is no doubt in his mind that the readers of his maps will prefer clear labels over accurate coastlines.

So, in this respect at least, al-Khwārazmī's project has failed. Al-Khwārazmī lived at the height of the translation movement, when Greek science was imported wholesale in the nascent Islamic culture. The shadow of the rationalist world view espoused by the caliph al-Maʾmūn loomed large, even if we have no proof that al-Khwārazmī actually worked at al-Maʾmūn's court. Over the following century, this world view lost some of its lustre. Greek knowledge was still valued, but was now adapted, rebranded and critically assessed. Traditionalist theologians reacted to the challenges of Greek philosophy by creating a theology that was more distinctively Islamic. A new generation of Muslim map-makers, led by the tenth-century map-maker al-Iṣṭakhrī, would create maps that put the civilization of Islam at the forefront of their attention.

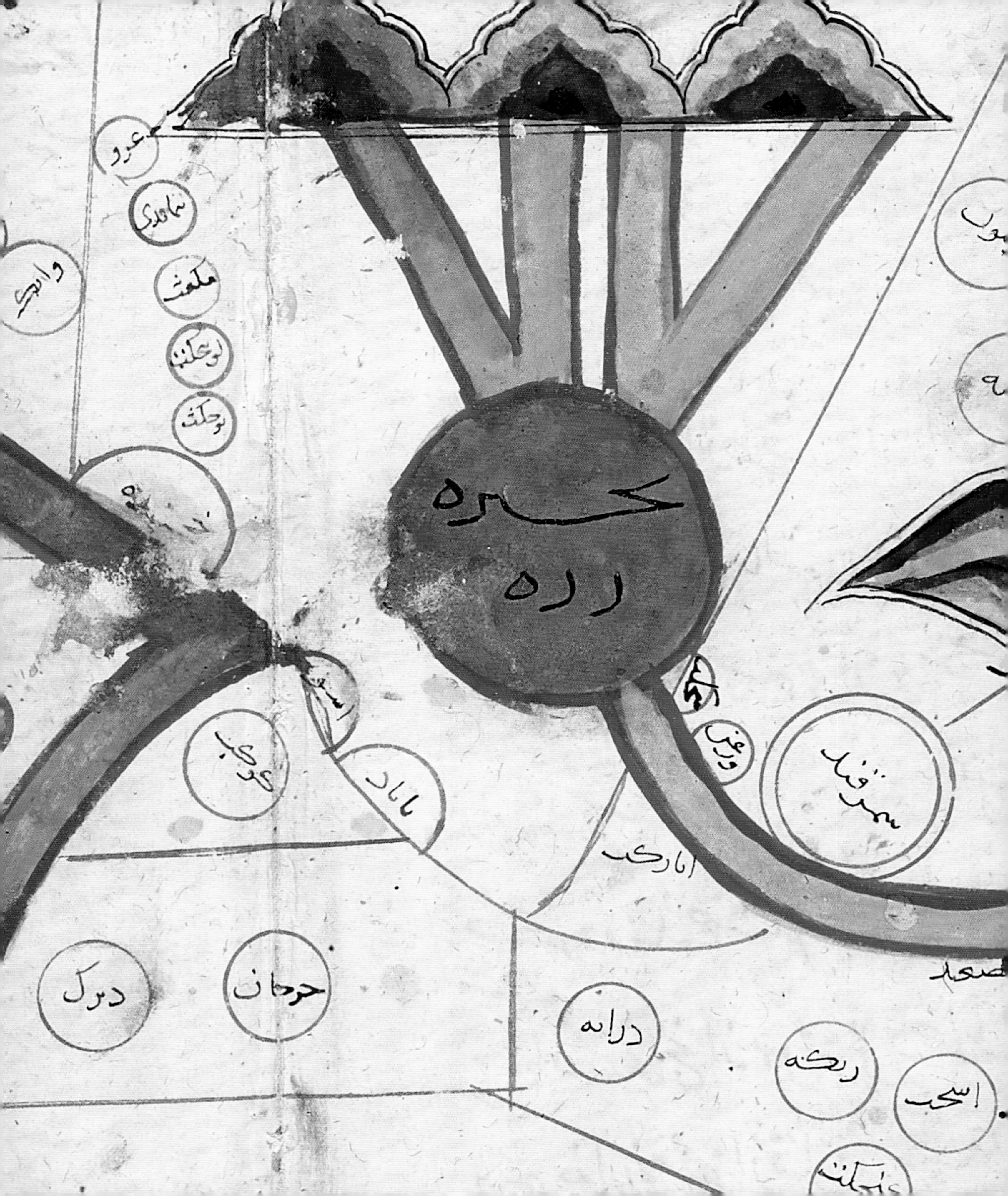
بحيره
زره
درك

CHAPTER 2

A World of Islam in Circles and Lines

Map of Syria and Palestine from al-Iṣṭakhrī's *Book of Routes and Kingdoms*, copied 1272. South is at the top. Bodleian Library, University of Oxford, MS. Ouseley 373, fol. 34r.

THE MOST SUCCESSFUL MAP-MAKER in Islamic history is someone we know hardly anything about, apart from what his name, his geographical work and his maps tell us. Abū Isḥāq al-Iṣṭakhrī, also known as al-Fārisī, was probably born in the ancient town of Iṣṭakhr in southern Iran towards the end of the ninth century. All we know about his life is that he passed through Basra in 935 or 936, and that sometime later he met a younger geographer called Ibn Ḥawqal to compare notes and maps of eastern and westerns regions of the nascent Islamic world. We do not know where their meeting took place. Both had travelled widely in an Islamic world that extended from India to Spain and that had – in the century since the death of al-Khwārazmī – fragmented into a plethora of dynasties and states, most of whom paid only token allegiance to the Abbasid caliph of Baghdad.

Al-Iṣṭakhrī's great legacy is his treatise on geography, known today as the *Book of Routes and Kingdoms*, a set of maps accompanied by geographical texts.[1] The original tenth-century manuscript has not survived; we only have later copies so we do not know exactly how al-Iṣṭakhrī wanted his maps to look. The earliest dated copy of his work, preserved today in the Gotha Research Library in Germany, was made at the end of the twelfth century. But in the following centuries – through the Late Middle Ages and the early modern period – under the Ottomans and the Mughals, scores of other copies of al-Iṣṭakhrī's work followed. There are today probably as many as fifty known copies of his work, often in translations from the original Arabic into Persian and Ottoman Turkish. There are far more known copies of al-Iṣṭakhrī's maps than of any other medieval Muslim map-maker. The hands of individual copyists introduced some variations into this corpus, and each manuscript of the the *Book of Routes and Kingdoms* is a work of art in its own right. Yet, despite these differences, they all clearly originate in one ingeniously simple design, which is distinctively al-Iṣṭakhrī's. The lingering popularity of his tenth-century maps – despite the ready availability of early modern European mapping technologies – suggests that they resonated with their Muslim audiences in ways that went well beyond their practical or scientific value. Indeed, al-Iṣṭakhrī's maps, more than any others, acquired a symbolic value as the iconic visualization of the interconnectivity and unity of the Islamic world.[2]

At the heart of his posthumous success is al-Iṣṭakhrī's revolutionary approach to map-making, which was the complete opposite to that of al-Khwārazmī. Al-Iṣṭakhrī eschewed any reference to mathematical geography and to coordinates of latitude and longitude, and dropped nearly all reliance on Ptolemy and Greek

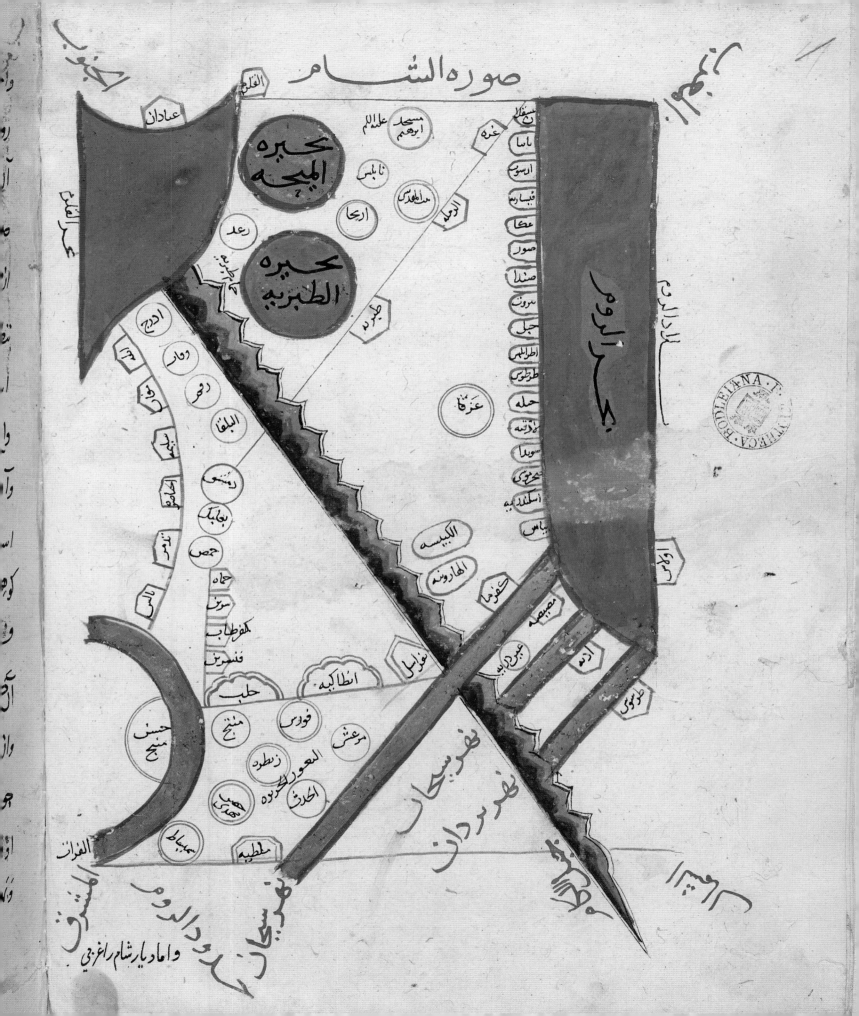

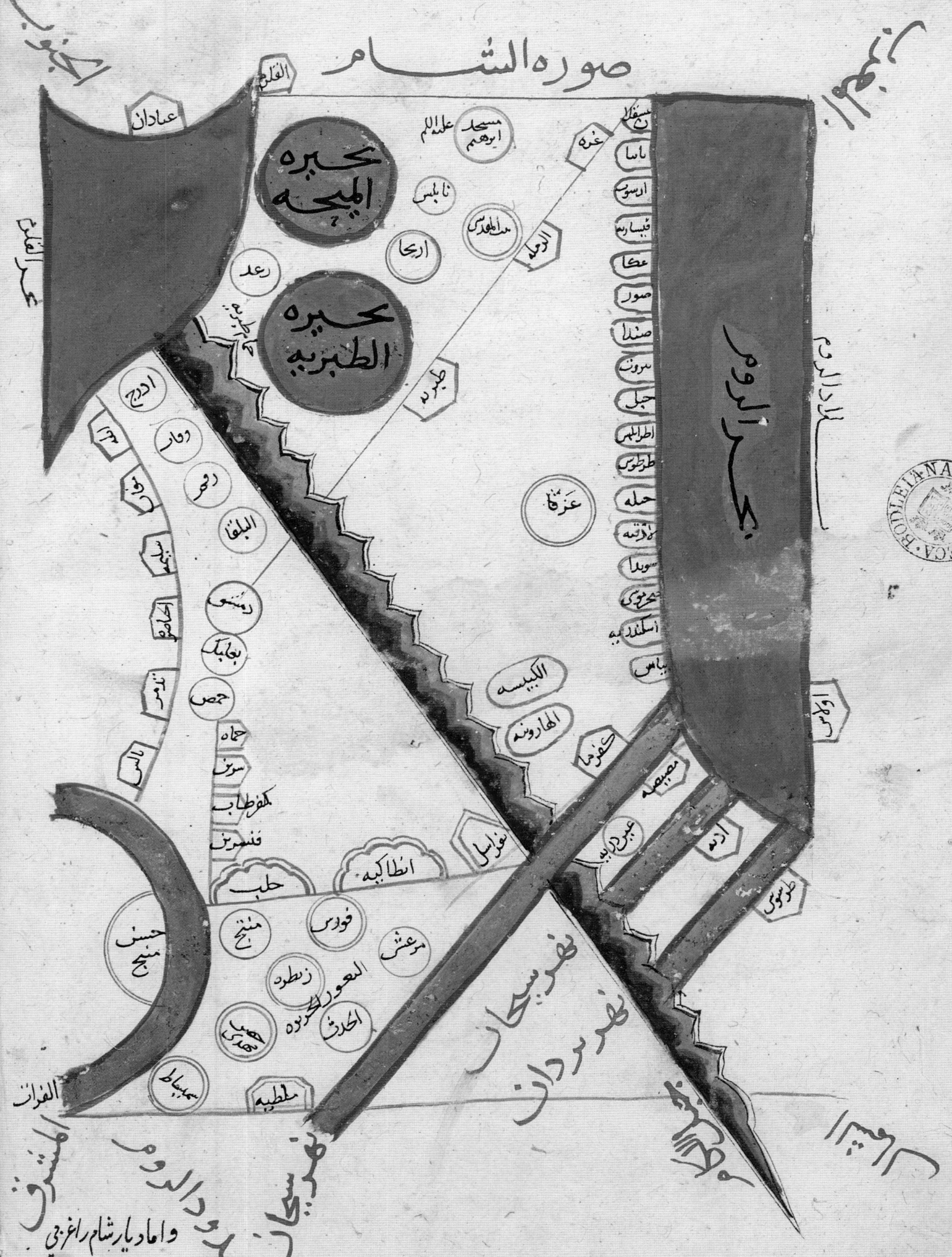

صوره الشام
بحر الروم
بحيره الطبريه
قيساريه
عكا
صور
صيدا
بيروت
اطرابلس
طرسوس
اذنه
مصيصه
الكنيسه
الهارونيه
انطاكيه
حلب
قنسرين
حماه
حمص
بعلبك
دمشق
البلقا
اريحا
نابلس
بيت المقدس
الرمله
غزه
طبريه
منبج
مرعش
زبطره
الحدث
ملطيه
سميساط
الفرات

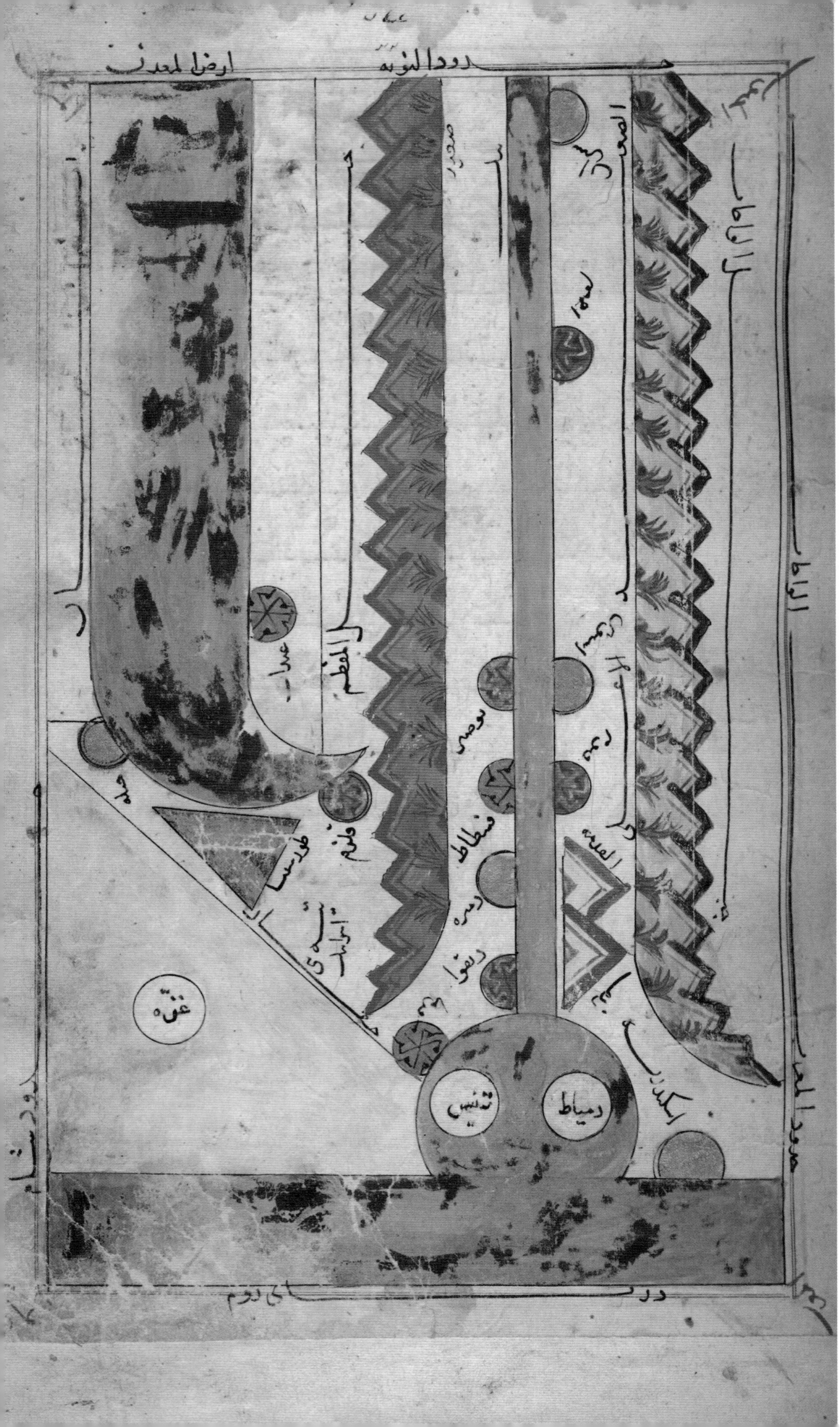

Map of Egypt from al-Iṣṭakhrī's *Book of Routes and Kingdoms*, copied 1306. South is at the top. The Nasser D. Khalili Collection of Islamic Art. Copyright Family Trust, MSS. 972, fol. 20a.

nomenclature. He also turned his gaze away from the universal and global outlook of his predecessor. Instead, his focus was the provinces of the Islamic world – that is, those regions governed by Muslim rulers. His work consists of twenty regional maps, a veritable atlas of Islam. Each regional map represents, in a trademark geometric design, land routes that are useful to commerce and pilgrimage. These itineraries are presented as straight lines on which stopping stations are indicated at roughly equal distances from each other, with a conscious, intentional disregard of scale. The non-Muslim lands excluded from the regional maps are included only in a circular world map, placed – as in a modern atlas – at the beginning of the book, with the aim of placing the regional maps that follow in relation to each other.

As an example of al-Iṣṭakhrī's regional maps, let us look first at his map of Syria and Palestine (p. 41). This map is preserved in a Persian translation, copied in 1272 and held in the Bodleian Library. The map is generally oriented to the south, indicated in the upper left corner. The green band on the right represents the Mediterranean, while the two equal-sized green circles on the top left represent the Sea of Galilee and the Dead Sea. The diagonal band indicates a series of mountain ranges than run from the Anti-Taurus in Anatolia, seen at the bottom right, to the Red Sea at the top left. The mountain range gives rise to rivers, which are indicated by straight blue bands. Along the Mediterranean coast we find a sequence of named towns, from Ascalon in the south to Tarsus in Anatolia. The thin straight lines connecting the inland cities indicate the major trade routes of the time. One such itinerary runs from Gaza in the top right to al-Ramla and Tiberias, then cuts through the mountain range to reach Damascus.

The most striking aspect of this map is its simple, abstract geometric design. Lines are either straight or curved, rivers represented as parallel lines and lakes are perfect circles. Towns are shown as squares, circles or other geometrically defined forms. The simple colour coding, with red for itineraries, and green and blue for salt and fresh water respectively, adds to the legibility of the map (although the copyist of this manuscript has wrongly coloured the Sea of Galilee in green, not realizing it is a freshwater lake). The line work is very stylized, the abstraction extreme; there is no attempt here to capture either scale or direction.

The disregard for mathematical geography presents a stark contrast to al-Khwārazmī's map of the Nile described in Chapter 1. For the sake of comparison, consider al-Iṣṭakhrī's map of Egypt (opposite), taken from a copy made in 1306–7 and held in the Nasser D. Khalili Collection of Islamic Art. First, we notice that al-Iṣṭakhrī's map is a map of Egypt and not of the Nile, and that it emphasizes the urban centres that the river connects rather than the actual bends and curves of the river path. This map has a richer colour palette, but the diagram remains distinctively abstract and schematic, with no attempt to represent the actual course of the Nile. Al-Iṣṭakhrī got rid of all the Hellenistic features that characterized al-Khwārazmī's map, and that testified to a heavy reliance on Ptolemy and pre-Islamic knowledge. The Mountain of the Moon and the lake system at the sources of the Nile, all derived from Hellenistic lore, are not shown in this map. Nor do we see any of the clime lines that linked al-Khwārazmī's map to mathematical geography. The Nile flows in a straight line from south to north, with a parallel mountain range on each side. Rather than ending with a forked delta, the river ends with a semicircular bay, in which lie the parallel, oversized islands of Tinnīs and Damietta. The Red Sea is shown on the left as a stylized bird's head, with the port of Qulzum (present-day Suez) at the point of the beak.

The Power of Abstract Geometry

Al-Iṣṭakhrī's abstract maps have been called primitive and naïve, but their abstraction is by design and not a result of ignorance or ineptness. As shown by Emilie Savage-Smith, the primary purpose of the geometric design was as an aid to memory, a means of imposing order on complex material.[3] The principle underlying al-Iṣṭakhrī's ingenious map-making is the same principle that underlies the iconic London Underground map, designed by Harry Beck (opposite). Like al-Iṣṭakhrī, Beck chose to simplify the map so that the railway lines appear as straight lines, and the stations as equidistant from each other, distorting the physical geography for the sake of legibility and accessibility. Precisely because Beck intentionally eschewed coordinates and an accurate representation of physical reality in order to provide an easily remembered way of sorting out complex information, his design has been immensely successful and popular since its introduction in 1933.

Like the London Tube map, al-Iṣṭakhrī's maps of Syria and Egypt served a practical purpose, with the caravan routes and their stations as the dominant feature. The source for the information on the maps must have come from travellers originally, but al-Iṣṭakhrī then transformed the data he collected to a visually accessible medium. Of course, these maps of Syria and Egypt were not pocket-sized objects taken on the road. They formed part of a longer treatise, and

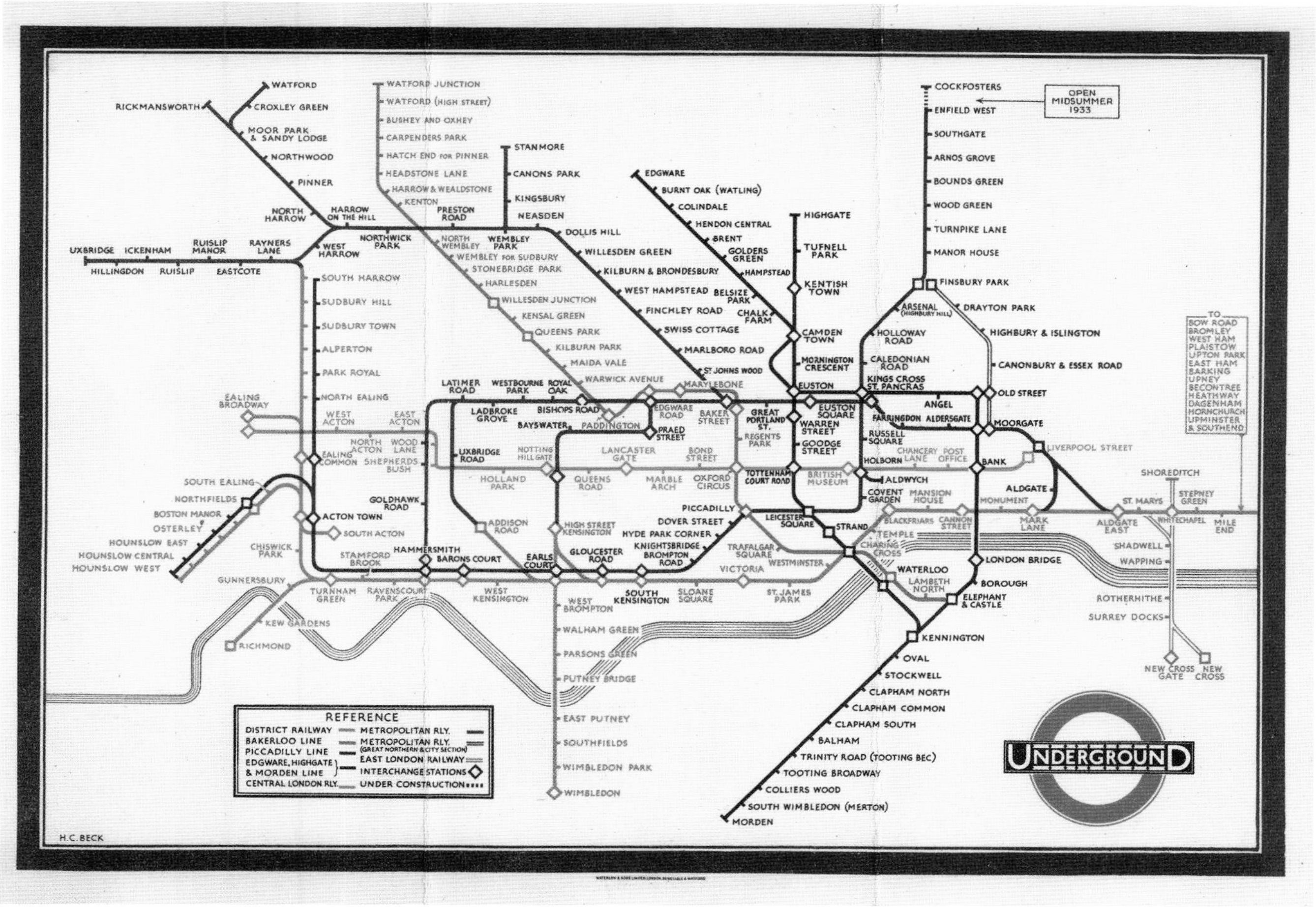

London Underground map, designed by Harry Beck, 1933. © TfL from the London Transport Museum collection.

each map was attached to a text about the human and physical geography of the region. But – and it may be useful to think of al-Iṣṭakhrī's work as the medieval equivalent of a modern atlas – the maps allowed the planning of journeys in a way that was not otherwise possible: by creating relationships between data in a way that narrative linear text could not, they enabled the map-reader to grasp the interchanges between routes.

Al-Iṣṭakhrī's geometric design also chimed with shifting tastes in art. During the first century of Islam, decorations on buildings made by the rulers of the new empire were often a variation on late antique and Hellenistic iconography, as shown in the mosaics covering the Dome of the Rock and in the rich frescos of the desert palaces of the Umayyad caliphs dating from the middle of the eighth century. But by al-Iṣṭakhrī's time the Abbasid ruling elites had begun to develop more distinctively Islamic modes of artistic expression. Whether in woodcarvings on palace walls or in new techniques of glazed pottery, artefacts from the ninth and tenth centuries showed a growing emphasis on repeated geometric patterns that created a complex web of stars, circles and other simple geometric shapes. This was recognized by Abbasid elites as both refined and specifically Islamic artistic representation, and when al-Iṣṭakhrī decided how to represent the world he turned to the same visual language.

Beyond the abstract geometry, these maps of Syria and Egypt also emphasize the bounded and separate character of these regions. The maps and the regions they depict are bounded by red or black lines at the edges of the map, and by the square form of the page. These maps reinforce what one modern scholar called a regional 'category of belonging'.[4] When al-Iṣṭakhrī was producing his maps in the tenth century, each of the regions of the Islamic world had already established some degree of distinct identity. These identities partly corresponded to the administrative and fiscal boundaries established by the Abbasid empire, but did not depend upon them. For example, the map of Syria and Palestine is followed by a textual description that explains further administrative subdivisions, such as the fiscal districts of Palestine (Filastin), Damascus and Jordan. But the map itself has no trace of these fiscal units. The local Syrian identity that emerges from the map does not hinge on a provincial governor or on tax collections, but appears to derive its integral character from the network of urban centres and trade routes.

While al-Iṣṭakhrī recognizes, bounds and reinforces the distinct identity of Syria and Egypt, he also presents these regions as the building blocks that together constitute the unity of the world of Islam. The regional maps all follow the same uniform geometric design and colour coding, and each map is connected to the regions next to it. The viewer of the map of Syria is made aware, through notations on the edges of the page, that they can continue to plan a mental journey through the lands of Islam by turning to the maps of the neighbouring regions elsewhere in the treatise, such as those of Iraq and the Arabian Peninsula. But there are no political divisions here, no indication of local potentates or autonomous sultans and no images of enthroned kings such as are found in later European portolan charts. The maps do not even give special attention to Baghdad as the capital of the Abbasid empire and the seat of the caliphs. The focus in al-Iṣṭakhrī's maps is on trade routes, and the Islamic world is envisaged as a chain of cities, a network of communication rather than a collection of rulers.

A World of Islam

Perhaps the greatest achievement of al-Iṣṭakhrī's maps is that they *created* the world of Islam as much as they reflected it. With the rapid fragmentation of the Abbasid empire, emergent Muslim communities were no longer connected to each other through a centralized political order. Instead, the unity of Islam was

Map of the Arabian Peninsula from al-Iṣṭakhrī's *Book of Routes and Kingdoms*, copied 1272. South is at the top.
Bodleian Library, University of Oxford, MS. Ouseley 373, fol. 4.

صورت دیار عرب

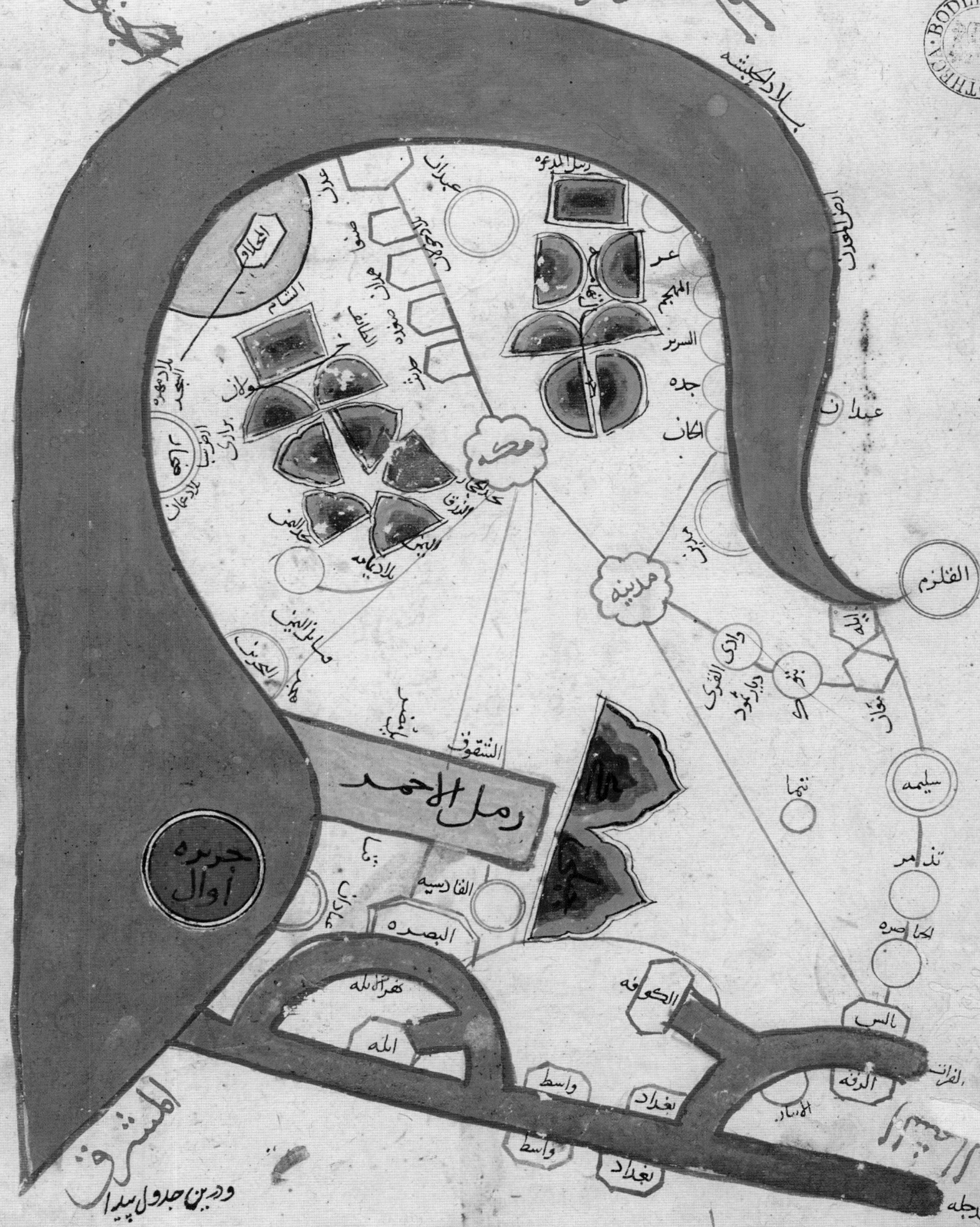

البحر المحيط
المشرق
ساكن البربر
تاهرت
سطيف
القيروان
زويلة
برقة
المهدية
حد مصر
بحر الروم
جزيرة صقلية
قرطبة
وادي الحجارة
افرنجة

Map of Muslim Spain and North Africa from al-Iṣṭakhrī's *Book of Routes and Kingdoms*, copied 1272. The Atlantic Ocean in the west is at the top of the map. Bodleian Library, University of Oxford, MS. Ouseley 373, fol. 23.

now preserved through the trade routes that connected urban communities of Muslim scholars and merchants, where local communities each had their own sense of regional identity but were all interconnected within a wider Islamic project. Al-Iṣṭakhrī's maps engaged with and responded to this polycentric and interconnected Islamic world, stretching from the Iberian Peninsula to the Indus Valley. By allowing many readers to visualize their home town as part of that Islamic world, al-Iṣṭakhrī may have helped to bring about that world.

In this new polycentric world of Islam, primacy of place was accorded to the Arabian Peninsula, and to the twin holy cities of Mecca and Medina. All copies of al-Iṣṭakhrī's work always place the map of the Arabian Peninsula before all other regional maps. The map of the Arabian Peninsula in the Bodleian copy (p. 47) is clearly dominated by the pilgrimage routes leading towards Mecca and Medina. The two cities are indicated prominently by large flower-shaped icons, from which itineraries radiate in all directions. The actual geography of the Arabian Peninsula is distorted, with Mecca and Medina placed at its centre, not in its western part as they are in reality. There are other features of interest in this map. To the west, the Red Sea again takes a stylized bird-shaped pattern. In the lower, northern, half of the map, the peninsula is separated from Iraq by a light purple band of 'red sand' stretching from the coast to the foot of a prominent twin-peaked mountain range north of Basra. Baghdad is shown too, right at the bottom, straddling both banks of the blue band of the Tigris. But the imperial centre is not given any visual importance here, being merely another link in the chain of cities that connect the different regions of the Islamic world.

Following the map of the Arabian Peninsula, al-Iṣṭakhrī's maps of the regions of the Islamic world usually start in the west and then proceed eastwards until they reach Central Asia. The first map in this sequence, representing the westernmost areas of the world of Islam in the tenth century, is that of Muslim Spain and North Africa (opposite). In this map, west is at the top-right corner. The large, perfectly shaped semicircle on the right-hand side represents Muslim Spain, with its capital, Córdoba, prominently marked as a large octagon at the centre of the peninsula – again, distorting the actual geography in order to render it more accessible and legible. From this central node, straight red lines indicating itineraries radiate in all directions. North Africa is shown on the other side of the green Mediterranean. One row of towns on the Mediterranean coast of North Africa is followed by a parallel itinerary inland. Further to the south, on the left of the map, the light purple area represents the sands of the Sahara.

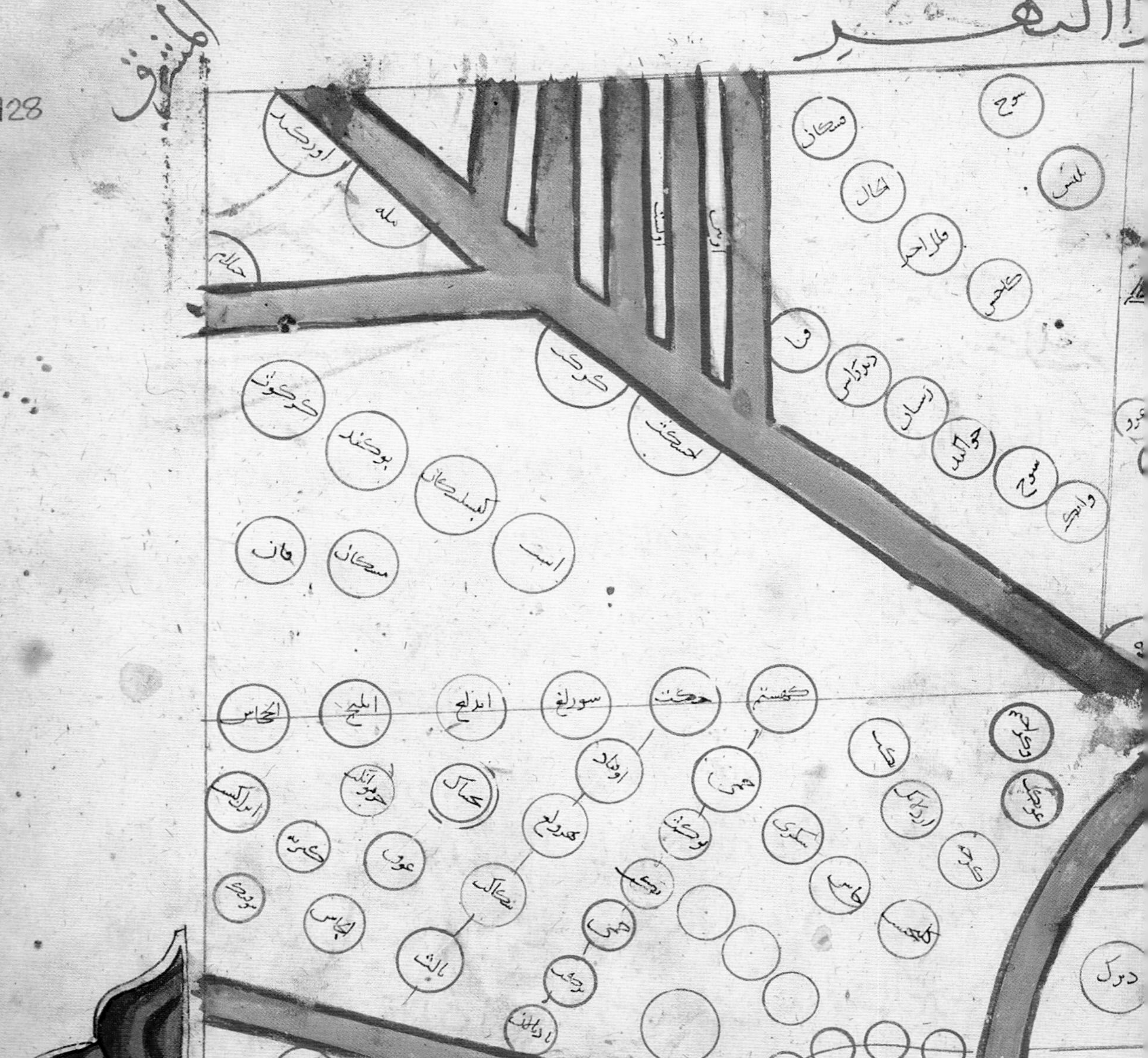
النهر
المشرق
حدود الترك

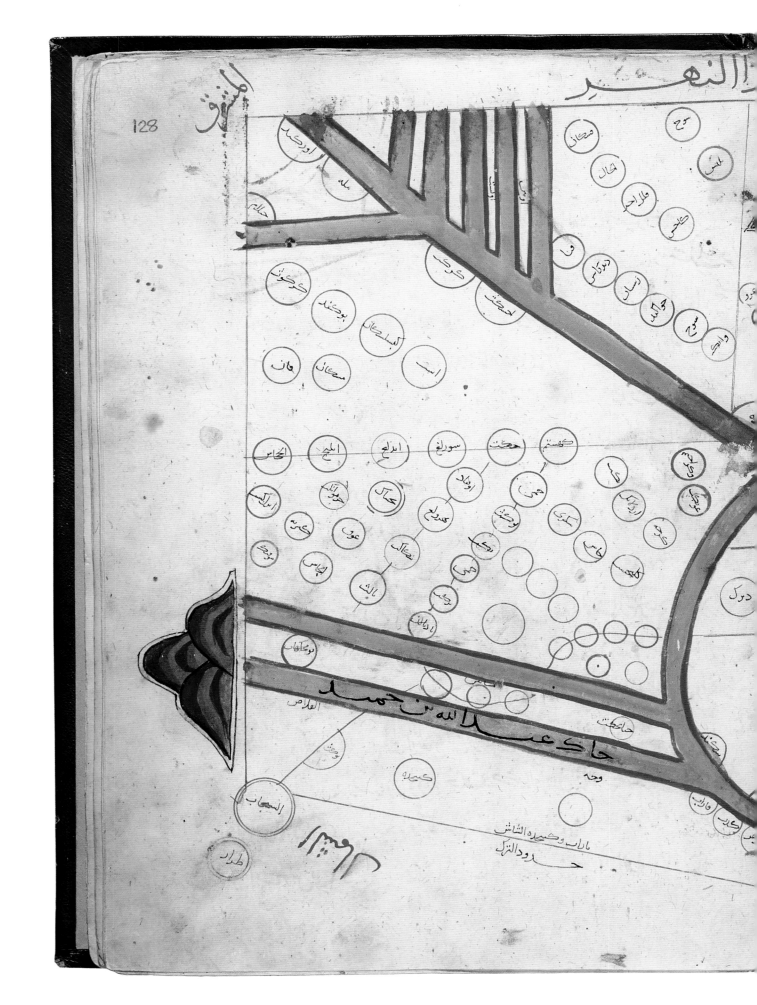

هذه صورة م
بحيره زره
بحيره خوارزم
الجنوب
سمرقند

Although this is a map that has the Mediterranean at its centre, al-Iṣṭakhrī's interest is primarily in the land, with little consideration for maritime space. Within the green of the sea, the large bright red circle is the island of Sicily, which was at that time the most important Muslim outpost in the central Mediterranean. The purple mountain above Sicily is the Rock of Gibraltar, awkwardly placed inside the Mediterranean Sea and not as part of the Iberian Peninsula. The straits at the top of the map lead to the Atlantic, here called the Ocean, or the Encompassing Sea. There is no attempt to reproduce actual coastlines on the islands or on the mainland. All the itineraries are on land rather than across the sea. In reality, ships could and did cross the short distance between Iberia and North Africa relatively quickly and easily, so that the two regions were joined together in their history and their identity. Al-Iṣṭakhrī acknowledges this shared identity by bringing Muslim Spain and Muslim North Africa together on the map and in the text that follows, but there is no explicit indication of maritime itineraries or of gulfs, bays and ports – on the whole, the sea appears to be alien territory.

Non-Muslim areas are excluded from the web of itineraries that form the world of Islam; they are, literally, off the map. The line that cuts through the Iberian Peninsula separates Islamic Spain from the lands of the Galicians in the north-west. The Galicians can be reached by following the red itinerary lines that radiate north from Córdoba, but their towns have no names and their lands are cut off from the network that makes Islamic Spain a distinct geographic and social unit. The Italian Peninsula is not shown at all, despite its obvious physical proximity to Muslim Sicily. The purple sands of the Sahara separate North Africa from the 'Land of the Blacks', and the trans-Saharan trade routes that brought gold and slaves to the shores of the Mediterranean are not shown. Al-Iṣṭakhrī set out not only to represent the world of Islam but also to define it as a self-sustaining whole that stood apart from those lands in which Islamic rule had not been established.

A native of Iran, al-Iṣṭakhrī was more familiar with and more at ease at depicting the eastern, Persian-speaking parts of the Islamic world. The last map in the sequence of the regional maps is a map of Mā Warā al-Nahr – literally, 'Lands beyond the river'. The river in question is the Amu Darya, known to the Greeks as the Oxus, and the region depicted here roughly corresponds to the modern states of Uzbekistan and Turkmenistan (previous spread). The map is oriented to the south-east. The large green circle on the bottom right, to the north-west, is the Aral Sea, and the two rivers that feed into it are the twin great Central

(previous spread) Map of Mā Warā al-Nahr (Iran and the Aral Sea), from al-Iṣṭakhrī's *Book of Routes and Kingdoms*, copied 1272. The map is oriented to the south-east. The large green circle on the bottom right is the Aral Sea. Bodleian Library, University of Oxford, MS. Ouseley 373, fol. 128.

Asian rivers, the Amu Darya coming from east and the Syr Darya from the north. The courses of the rivers are, as usual, distinctively stylized, with tributaries shown as straight lines or as half-circles. Bukhara, the most important city in the region at the time, is shown as two purple circles straddling the banks of the Zeravshan River, near its confluence with the Amu Darya. Further upstream on the Zeravshan is Samarkand, another key station on the caravan routes, which is circled with a double red line below a three-peaked purple mountain.

This map of Islamic Central Asia, covering two folios of the manuscript, is overflowing with towns and stopping stations, which are usually connected to each other by straight or curved lines. While the narrower map of the Islamic West is bounded by the sea on the one hand and by the lands of the unbelievers on the other, this map conveys the expanse of the Central Asian steppe and its criss-crossing, decentralized caravan routes. The wealth of information here probably came from a variety of sources, including al-Iṣṭakhrī's own travels in these regions. We know that he relied heavily on the geographical work of a scholar from a previous generation named al-Balkhī, who had worked in Bukhara, then the capital of the important Samanid dynasty. Thus, al-Iṣṭakhrī was building on knowledge acquired in new centres of learning that had sprung up along the route to China, at a time when the centre of gravity of the Islamic world was shifting eastwards. Centuries after his death, his focus on the eastern Islamic lands had a special appeal in Iran. It was there that three Persian translations of al-Iṣṭakhrī were produced during the thirteenth century. A unique copy of one of these translations ended up in the Bodleian Library in Oxford.

As in a modern atlas, al-Iṣṭakhrī opens his treatise with a stylized map of the world that precedes all the regional maps (pp. 54–5). The main aim of the world map is to allow the reader to place the different regional maps in relation to each other. This world map is completely different in its design from the world maps associated with mathematical geography that were discussed in Chapter 1; the only continuity between them is the convention of placing south at the top. A circular rather than rectangular map, this shows all the continents known at the time, whether Muslim or non-Muslim, inhabited or uninhabited. The outside ring enclosing the map is the Encompassing Sea, and al-Iṣṭakhrī assumed, like all medieval scholars before Columbus, that the other hemisphere is completely covered by water. Europe is the smallest continent, represented as a triangular island in the bottom right, separated from Asia by the Black Sea. The three prominent Mediterranean islands are Cyprus, Crete and Sicily. Africa is

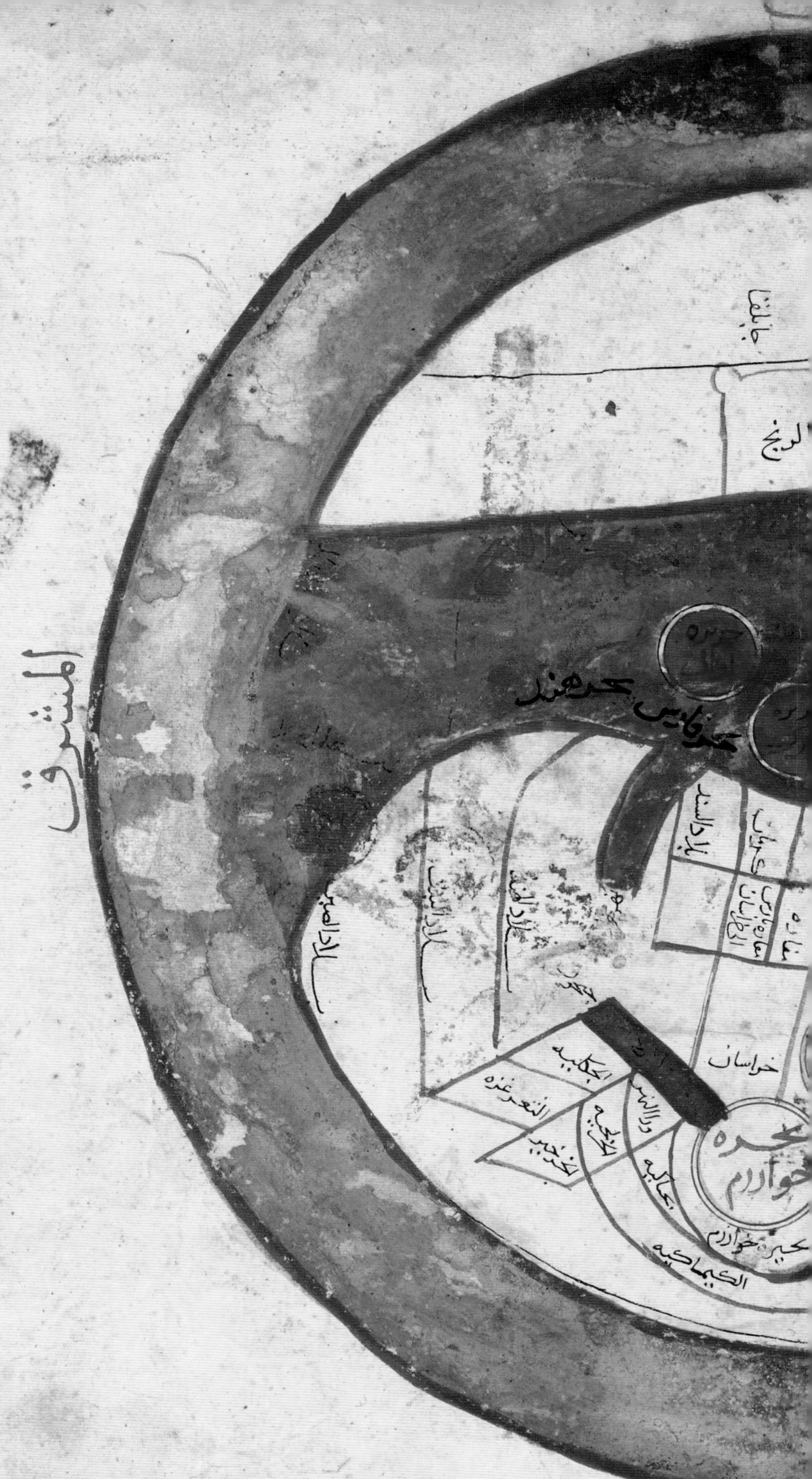
المشرق
بحر فارس بحر هند
بلاد السند
بلاد الهند
بلاد الصين
خراسان
التغزغزه
خرخير
بحيره خوارزم
الكيماكيه

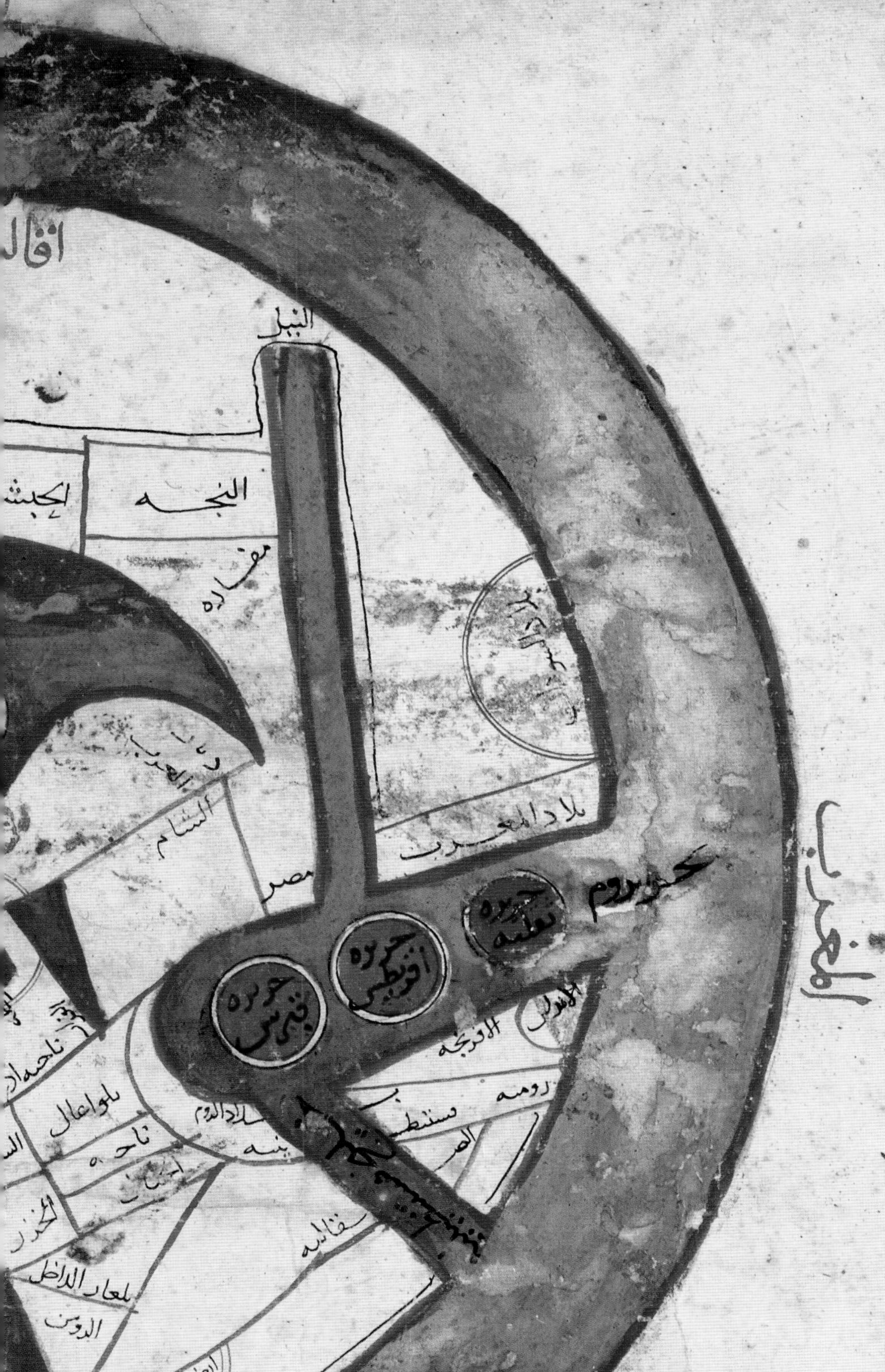

النيل
النجه
مفازه
بلاد المغرب
مصر
الشام
بحر روم
جزيره صقليه
جزيره اقريطش
جزيره قبرس
الافرنجه
رومه
الروس
المغرب

bisected by the Nile, represented as a straight channel, and the Red Sea is again recognizable by its stylized beak-like shape. The Indian Ocean is open on the left to the Encompassing Sea. The three perfect large circles in the middle of the Indian Ocean are, surprisingly, representations of three tiny strategic islands in the Persian Gulf, not far from al-Iṣṭakhrī's home town in southern Iran.

(previous spread) Map of the world from al-Iṣṭakhrī's *Book of Routes and Kingdoms*, copied 1272. South is at the top. Bodleian Library, University of Oxford, MS. Ouseley 373, fols 2–3.

Al-Iṣṭakhrī's world map is divided into neat rectangles and other geometric shapes, mostly corresponding to the areas depicted in the regional maps later in the book. Here Egypt is boxed between the Nile and Syria (al-Shām), and the latter is found beneath a curved Arabian Peninsula. The region of Mā Warā al-Nahr is next to the Amu Darya, adjacent to the prominent Aral Sea, indicated by a double-lined circle. The lands of North Africa inhabit a square between the Nile and the Atlantic. They are separated from Islamic Spain (al-Andalus), which is shown as a circle at the westernmost sharp tip of Europe. Beyond the regions of the Islamic world, this world map guides the viewer through the lands of the unbelievers, without drawing any hard boundaries between what jurists termed the 'domain of Islam' and the 'domain of war'. In Europe, rectangles represent the lands of the Franks, the Byzantines and the Slavs. Pagan Turkic tribal groups such as the Oguz and Pechengs have their territories marked in the Central Asian steppe. China is in the far east, facing the non-Muslim people of the Zanj in East Africa.

The purpose of this world map is explained by al-Iṣṭakhrī in his introduction to the treatise:

> I made an image [*ṣūrah*] of the entire Earth surrounded by the Inaccessible Encompassing Sea, so that when a viewer looks at it he would know the location of each region we mentioned, how they link with each other and the size of each region in relation to the whole world. Then, when he views each region in detail he can refer to its location in this image. This map, where all the other regions are also represented, is not large enough to capture the width and length of each region, and its different forms – whether they are round, rectangular or triangular – as they are shown in the map of that region. On the world map I merely indicate the location of each region so that its place is known. This is followed by maps dedicated to each of the lands of the kingdom of Islam, where I show the form of the region, its cities and all that it is necessary to know about it.[5]

As al-Iṣṭakhrī explains, the purpose of this world map is to show the regions of the world in relation to each other. He sees no need to name individual cities, to

Map of the Maghreb, including the entire Mediterranean Sea, from Ibn Ḥawqal's *Of the Image of the World*, copied 1086. Topkapi Palace Museum, Istanbul, MS. 3346, copied in 1086/479H, fols 19a, 19b, 20a.

indicate mountains or to reproduce any realistic coastlines. The map is divided into basic geometric shapes, which are easily legible and easily reproduced by copyists. While it shows non-Muslim regions of the world to place them in their geographical context, the world map is fundamentally subordinate to the regional maps that follow. The regional maps, as al-Iṣṭakhrī explains, cover only the regions that are part of the kingdom of Islam (*mamlakat al-islām*), a neologism invented to capture the politically fragmented yet culturally united Islamic world that emerged after the decline of Baghdad.

Al-Iṣṭakhrī's revolutionary abstract maps had a long-lasting impact on later generations of map-makers. Of particular importance is a younger contemporary called Ibn Ḥawqal, a merchant as much as a scholar, who placed great importance on direct experience of the countries he described.[6] A native of Mosul in Iraq, Ibn Ḥawqal set out in 943 to travel in the western regions of the Islamic world. His *Kitāb Ṣūrat al-Arḍ* (*Of the Image of the World*), which was heavily influenced by al-Iṣṭakhrī's work, also consists of twenty maps of Muslim-ruled regions preceded by a diagrammatic circular world map. The earliest version, produced sometime between 965 and 969, reproduced most of al-Iṣṭakhrī's text and maps, only updating the chapters concerning North Africa and Muslim Spain. Overall, Ibn Ḥawqal was a disciple who was clearly influenced by al-Iṣṭakhrī's geometric design and who also eschewed any aspect of mathematical geography.

But when the two met (as mentioned at the beginning of this chapter), Ibn Ḥawqal criticized al-Iṣṭakhrī's maps of the western Islamic world, where he himself had travelled widely. In the latest version of his geography, written sometime after 979, he was tempted to move away from the extreme abstraction that characterized al-Iṣṭakhrī's maps. In one beautiful three-folio map, entitled 'Map of the Maghreb and Byzantium' (p. 57), Ibn Ḥawqal has opted for a much more recognizable, more elaborate, more detailed and less schematic Mediterranean Sea, with numerous peninsulas and islands indicated, and attention paid to both scale and orientation. As indicated in the title of the map, Ibn Ḥawqal also opted to include much more of the non-Muslim regions of this Mediterranean world.[7] Both Italy and the Peloponnesus are now present and easily identifiable, and there are numerous bays along the North African coast. The Mediterranean islands are no longer blown out of proportion and, while the Strait of Gibraltar is oversized, the Rock itself is correctly placed at the exit point to the Atlantic.

But, although Ibn Ḥawqal's map of the Mediterranean was more accurate than that of al-Iṣṭakhrī, and was evidently based on first-hand experience, his approach was far less successful. The map shown here comes from an early copy of Ibn Ḥawqal's treatise made in 1086, but very few later copies of this work have survived, and those that have tend to be less detailed and more schematic. It was al-Iṣṭakhrī's maps, with their strict geometric design, that proved to be much more popular in the long run. Ibn Ḥawqal's map may be more accurate, but in his drive for greater precision he lost some of the simplicity and accessibility of the original maps. He also lost the sharp exclusive focus on the Islamic world that defined al-Iṣṭakhrī's distinctive approach. Ibn Ḥawqal's image of the Mediterranean was both difficult to copy and too dense to be of use as a quick reference guide for pre-modern users.

Al-Iṣṭakhrī's simple design, however, continued to appeal to other map-makers, even as they gathered more accurate data about the shape of the world. A superbly insightful glimpse of this comes from the pen of another geographer and traveller, the Palestinian scholar al-Muqaddasī, who lived in the second half of the tenth century. Al-Muqaddasī structured his work as another 'atlas of Islam', with a world map followed by regional maps of the Islamic lands. While visiting the port of Aden, he approached a leading local merchant, whose ships travelled all over the Indian Ocean to India and to Egypt. Al-Muqaddasī asked the merchant to describe for him the contours of what he called the Persian Sea. In response, the merchant did not only explain the shape of the Indian Ocean in words, but also drew an impromptu map with his hand:

> He rubbed the sand with his palm and drew the sea on it. The sea did not have the shape of a shawl [*ṭaylasān*] or of a bird. He drew winding coastal features and many gulfs, then said: 'This is the shape of this sea, there is no other map [*ṣūrah*].'[8]

Al-Muqaddasī had before him a detailed map of the Indian Ocean made by the most well-informed person at the time, and this map did not at all resemble the maps made by al-Iṣṭakhrī. The curved and winding Indian Ocean shores were not depicted in simple geometric forms, and the Red Sea did not have the iconic shape of the bird that can be seen in al-Iṣṭakhrī's various maps. And yet, al-Muqaddasī proceeds to tell us, he decided not to use the image drawn by the merchant in his own work. Instead, he made his own map of the Indian Ocean intentionally simple and purposefully naïve (*sādij*). The map he produced (p. 61) leaves out all the bays and gulfs described by the Adeni merchant, retaining only the bird-shaped Red Sea, which was, he says, too well known and not disputed.[9] For al-Muqaddasī, as for al-Iṣṭakhrī before him, this naïve schematic style was a choice rather than a reflection of limited knowledge, a way to impose order and legibility on a complex physical reality.

The intended audience of these maps, as Ibn Ḥawqal says in his introduction, are 'princes and people of consequence, who are the only ones preoccupied with this kind of knowledge'.[10] Al-Muqaddasī tells us that he looked at maps of the Indian Ocean found in the treasuries of the emirs of Baghdad and of Khorasan, in eastern Iran. Ibn Ḥawqal dedicated one version of his geography to Prince Sayf al-Dawla of Aleppo. And yet, although they were intended for elite consumption, these maps were conceived as maps for practical use and not as items of entertainment or for contemplation. The earlier medieval copies of the works of these three geographers carry little decoration, nor do these maps convey a theological message about salvation: Mecca is foregrounded but is hardly the centre of the world. And, while the intended audience was the ruling elite, the information on the maps and in the attached prose sections is not primarily political but rather cultural and economic, suggesting that Ibn Ḥawqal had in mind scholars and merchants as much as men of the sword.

The earliest dated copy of al-Iṣṭakhrī's maps was made at the end of the twelfth century, and many more copies were made from the thirteenth century onwards. By this time, however, they had begun to serve a different purpose and to appeal to different audiences. The first translation into Persian appears to have been made for one of Genghis Khan's provincial governors in Central Asia. Al-Iṣṭakhrī's focus on the Islamic east must have been part of the appeal of the treatise in that historical context, where the Mongol conquest intensified the separate regional identity of the Muslim territories that came under direct Mongol rule. In the Middle East, al-Iṣṭakhrī's abstract world map was sometimes divorced from its original treatise, and adapted and incorporated into a new genre of illustrated *mirabilia* treatises, which eventually became very popular in the Ottoman world. Other adaptations of the world maps were sometimes added to chronicles as explicatory diagrams. Some of the late medieval copies were lavishly coloured but, nonetheless, continued to strictly adhere to the abstract design without attempting to add detail or to update the names of places and tribes. The tenth-century Islamic world of al-Iṣṭakhrī's day became frozen in time.

Al-Iṣṭakhrī's maps gradually became *objets d'art*, illuminations produced at great expense and decorated with gold.[11] They were copied throughout the eastern Islamic world, where the Persian translations were especially popular. Today, the majority of the surviving manuscripts of al-Iṣṭakhrī's work are in Persian. One of the latest copies, made in Kabul in the nineteenth century, is embellished with a naked female figure and fishes drawn from the miniature painting tradition, but the simple contours of the graphic design are retained and reinforced.[12] The early modern consumers of these maps appreciated al-Iṣṭakhrī's maps for their nostalgic value, for they hailed from what was perceived to have been an Islamic golden age. The images came to be mentally connected with the set of texts and ideas that formed the canon of Islam; they acquired the aura of immutability that was accorded to works of law and of theology. It also helped that al-Iṣṭakhrī's design was beautiful in its simplicity.

The story of the afterlife of al-Iṣṭakhrī's maps is as fascinating as the story of the original design. In recent years, the map historian Karen Pinto has identified a cluster of copies of al-Iṣṭakhrī's treatise commissioned by one of the greatest Ottoman sultans, Mehmed II, following his momentous conquest of Constantinople in 1453.[13] These copies were part of a project of Islamizing the newly conquered capital, which was quickly renamed Istanbul. The copies made in Istanbul were simple and unadorned, as shown by the world map here (p. 62). But all six were based on a lavishly executed manuscript, with lapis lazuli and gold pigment, that Mehmed had probably received as a gift a few years earlier from the ruler of the Iranian city of Tabriz.

The reproduction of al-Iṣṭakhrī's maps appears as a peculiar choice on the part of the sultan. Mehmed, whose image has been preserved for posterity through a famous portrait made by the Venetian painter Giovanni Bellini, is well known for his interest in European maps, which he used for planning his European campaigns. We also know that he owned and consulted one of the early Ptolemaic world maps made in Byzantium in the first half of the fifteenth century, maps that

Map of the Indian Ocean from al-Muqaddasī's *The Best Divisions for Knowledge of the Regions*, copied 1494. The map is broadly oriented west, with the Red Sea at the top. © bpk Bildagentur / Staatsbibliothek Berlin, Orientabteilung, MS. Sprenger 5 (Ar. 6034), fol. 4.

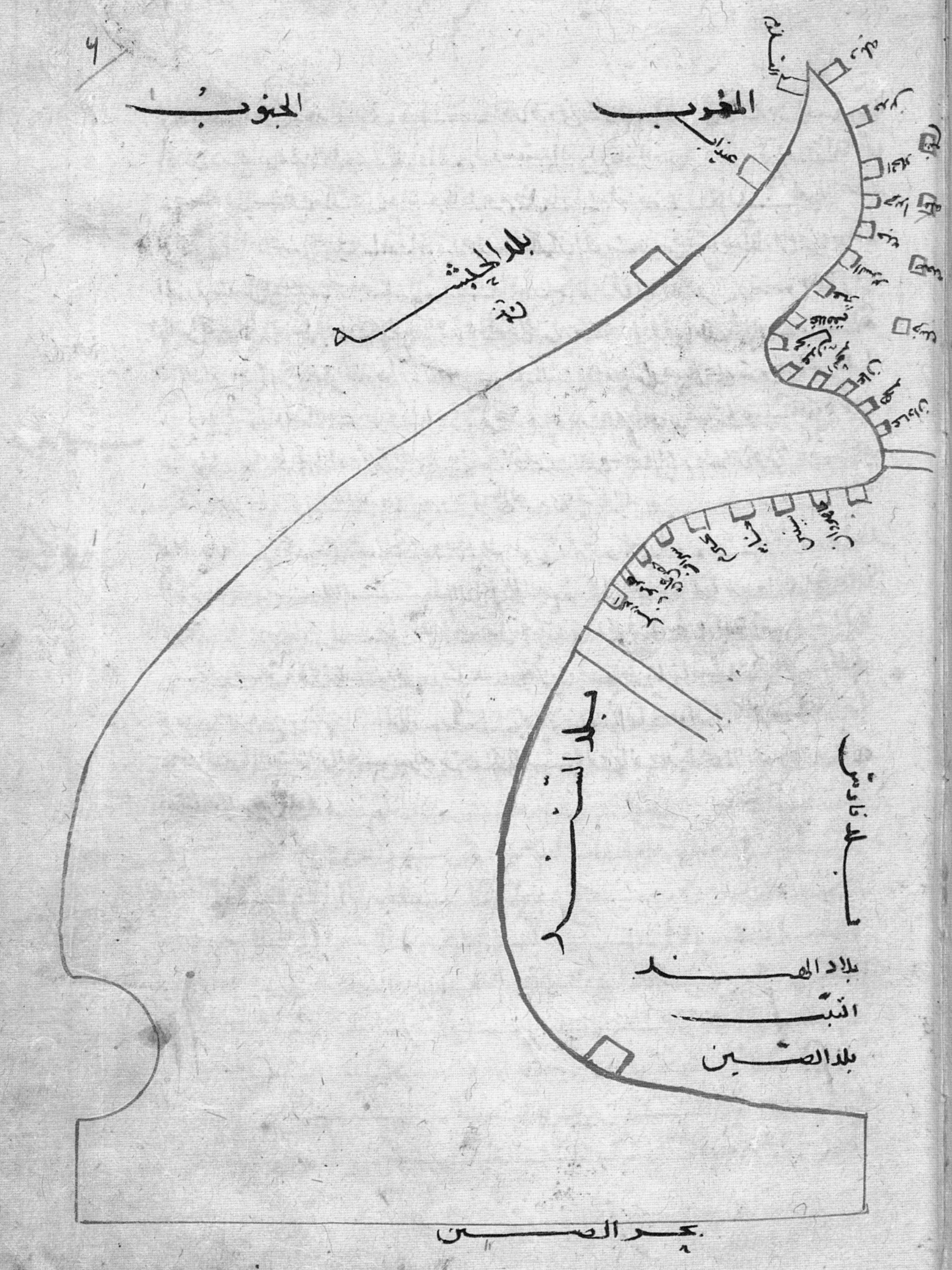

الجنوب
المغرب
بلاد الحبشة
بلاد السند
بلاد فارس
بلاد الهند
التبت
بلاد الصين
بحر الصين

صورة الكل

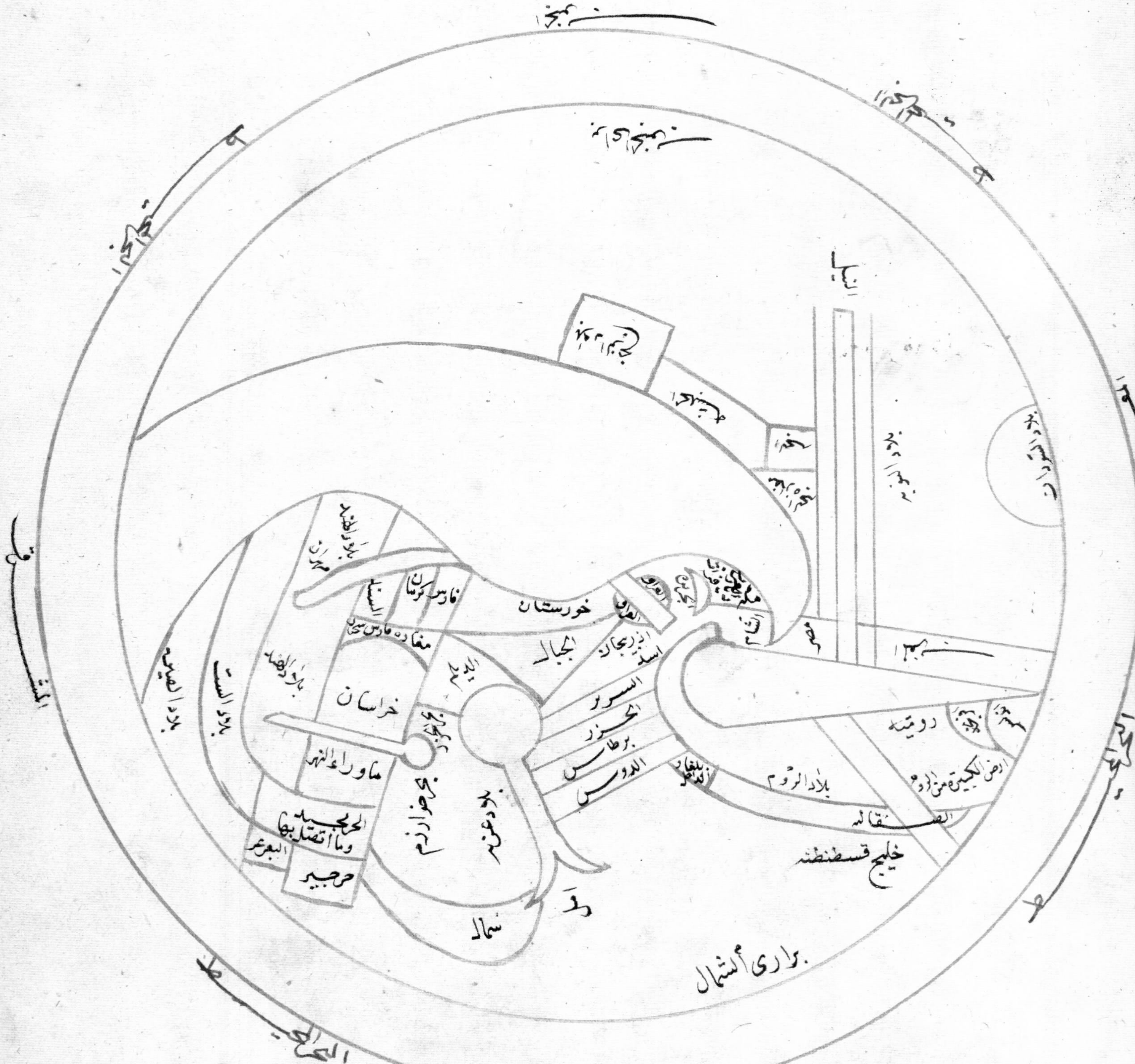

خورستان
الجبال
خراسان
ماوراء النهر
خوارزم
الديلم
السرير
الخزر
برطاس
الروس
مصر
روميه
بلاد الروم
الصقالبة
خليج قسطنطنية
شمال
براري الشمال
البحر المحيط
الشمال

Map of the world from al-Iṣṭakhrī's *Book of Routes and Kingdoms*, copied in Istanbul in 1523, from an earlier manuscript of 1473. South is at the top.

relied on mathematical geography and offered a closer approximation of physical realities. Residing in his new capital, Mehmed knew that al-Iṣṭakhrī's world map was wrong to show the Black Sea cutting off Europe from Asia and connecting to the Encompassing Sea. Nor was his own Ottoman dynasty mentioned on the map, where the now-defunct Byzantine empire was still highly visible.[14]

Ultimately, Sultan Mehmed chose to commission al-Iṣṭakhrī's set of archaic and schematic maps showing the world as it had been six centuries earlier because these maps held a symbolic value. They were a visual representation of the link with Islamic traditions and with Islamic history, somewhat comparable to the copying of the books of the Islamic prophetic traditions. These traditional maps were for public consumption, to be handled by religious scholars, unlike European maps, which were intended for private, courtly consumption. Mehmed ordered the six simplistic and unadorned copies of al-Iṣṭakhrī's work to be housed in the libraries of the new madrasas (institutions of learning) that he founded in Istanbul. By the late fifteenth century, the value of al-Iṣṭakhrī's maps no longer lay in their actual representation of the world but rather in their being romanticized mementos of the Islamic past.

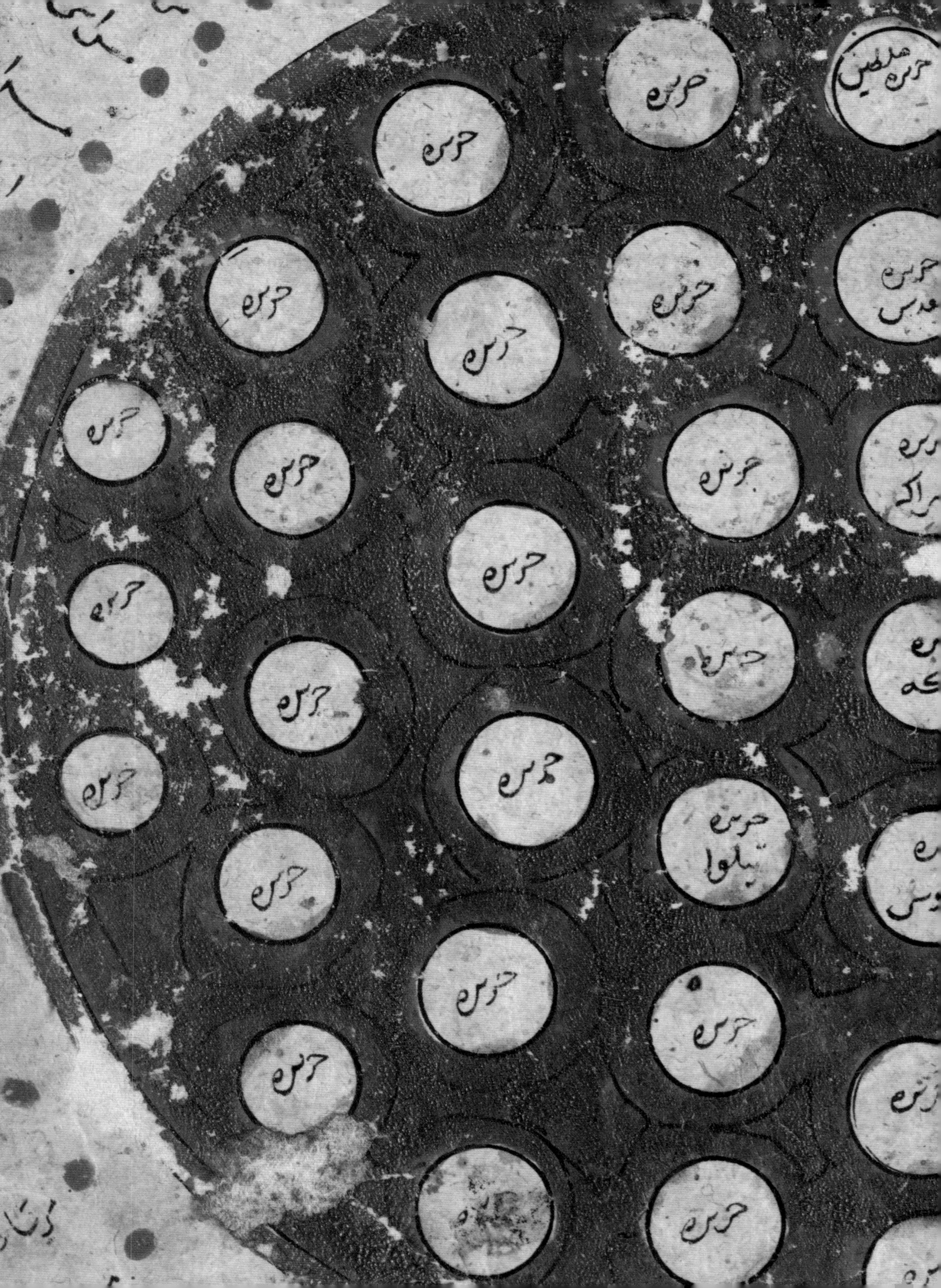
حرس
حرس
حرس
حرس
حرس بلوا

CHAPTER 3

The Mysterious *Book of Curiosities*

ON 10 OCTOBER 2000 a rather scruffy manuscript, bound in ill-fitting covers, went on sale in Christie's auction house in London. It contained a medieval Arabic treatise on the skies and the Earth by an unnamed author, accompanied by a series of strange images and maps unparalleled in any other medieval work. This treatise, whose Arabic title translates as the *Book of Curiosities of the Sciences and Marvels for the Eye*, turned out to be one of the most important discoveries in the history of cartography in recent decades. Virtually unknown to modern scholars before that public auction, it has since transformed our understanding of Islamic map-making.

One of the largest images in the manuscript was this curious map of an oval island (opposite), entitled 'Twelfth Chapter Presenting a Brief Description of the Largest Island in These Seas'. Everything about this map was clearly out of the ordinary. The subject matter of the map appeared to be an individual island, an uncommon subject in the corpus of Islamic maps that was known at the time. Within the island, the map appears to zoom in on a single round city, showing its walls and gates. It also shows distinctive twin towers at the top of the map and an onion-domed structure to the right of the walls, in an attempt to represent the distinctive architectural features of a medieval Islamic town pictorially.

The labels are badly corrupted, but a team of Oxford scholars began to decipher them one by one. The onion-domed building is called here the 'Palace of the sultan'. Each of the twin towers is labelled the 'Tower of the chain'. The mountain with the red cap at the bottom left is labelled the 'Mountain of Tilla' (جبل التلة), but this must be a copyist mistake for the 'Mountain of fire', meaning a volcano. As for the gates along the walls, one of the gates closest to the sea is labelled the 'Gate of the sea', while another one, called 'Gate of Shaghāth', is placed on the left-hand part of the walls. The Oxford team concluded that this must be the Gate of Saint Agatha, which still stands today in the old city of Palermo. Everything else fell into place. The volcano is Mount Etna, the 'seas' in question are the Mediterranean, and the 'Largest island in these seas' is, undoubtedly, Sicily.[1]

This depiction of Sicily surprises us in several respects. Most modern audiences do not immediately associate Sicily with the world of Islam. Yet Sicily was a major Muslim stronghold, a key outpost in the central Mediterranean until its conquest by the Normans in 1068. A millennium later, the impact of Islam is evident in many of the place names around the island, and even in the local dialect. We are also surprised by the oval form of the island. The triangular shape of Sicily

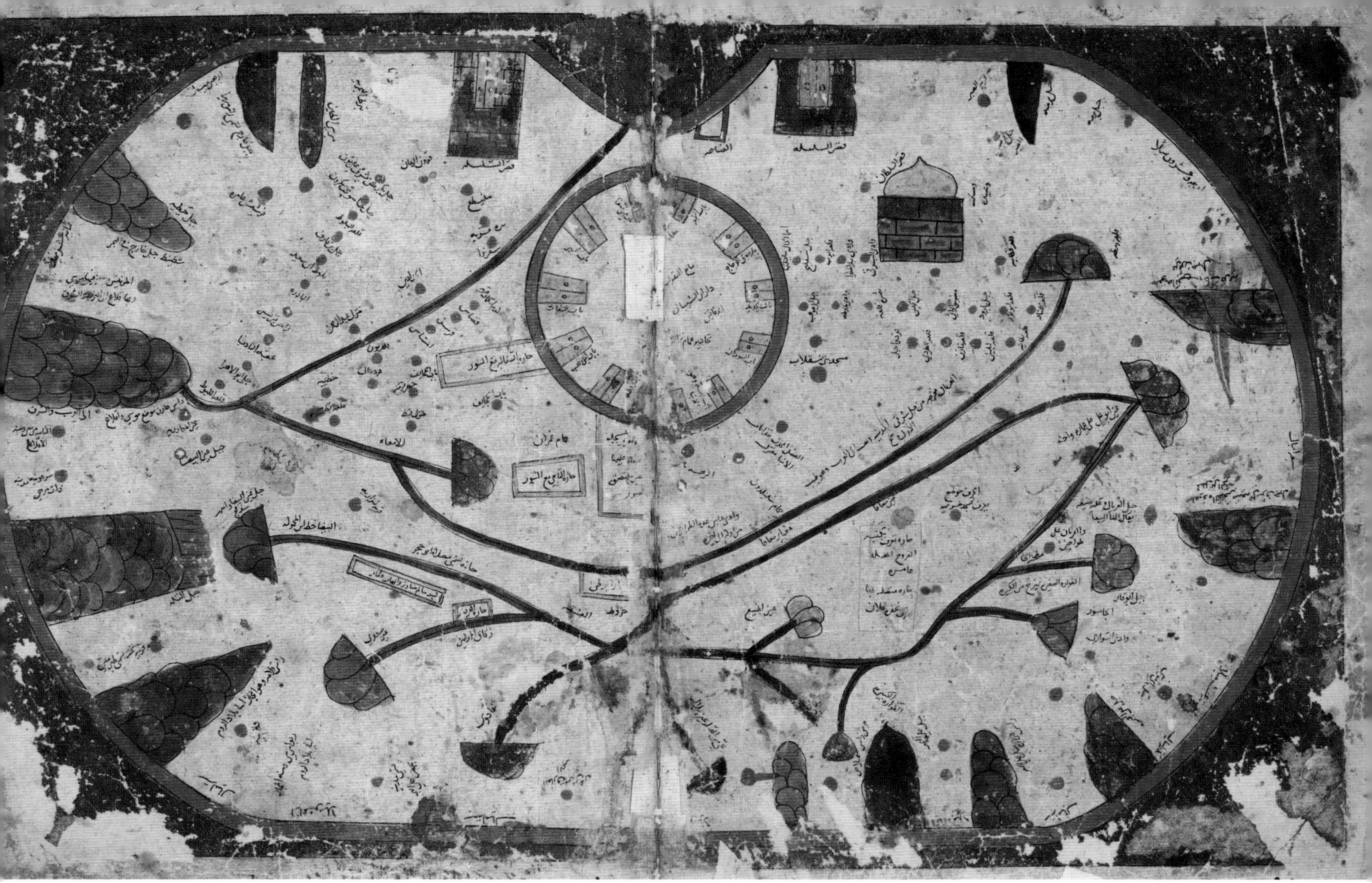

Map of Sicily from the eleventh-century *Book of Curiosities*, copied *c.*1200. Bodleian Library, University of Oxford, MS. Arab c. 90, fols 32b–33a.

is so iconic that we assume that every graphic representation would evoke that familiar shape. The map-maker chose not to do so. Like al-Iṣṭakhrī before him, he preferred geometric abstraction over actual coastlines.

Another curious feature of this map is that the walls of Palermo, the towers and the palace are bloated out of proportion and with no attention to scale in the same way as Manhattan is foregrounded in famous tourist maps of New York. We expect to see Sicily, but we actually zoom in on its capital. The V-shaped indentation at the top of the map represents the port of Palermo, flanked by twin towers from which a chain could be lowered to block access. The red box to the right of the harbour is the arsenal – the English term comes from the Arabic *dār al-ṣināʿah*, where galleys were built. The domed structure at the top right, labelled the 'Ruler's palace' (*qaṣr al-sulṭān*) represents the palatial compound of al-Khāliṣah, built in 937 to house the ruling dynasty. Today this area in the old city of Palermo is called La Kalsa.

Visually, the landscape of Palermo is dominated by the circular walls and eleven gates. In a text that precedes this map, the author informs us that Palermo originally took the form of a square, but that it had expanded to form a circle, and this is how he represents the city on the map. Inside the city walls are four labels not accompanied by any iconography. Two of these labels refer to intramural markets, one of herb sellers and the other of flour merchants. The two other labels are enigmatic references to the 'House of Ibn al-Shaybānī' and to the 'Baths of Nizār'.

There are also eight suburban quarters, which are indicated by rectangular boxes with thin yellow borders. The thinness of these yellow borders is a visual reminder that their defences are not as solid as those of the city itself. Two adjoining quarters on the bottom right ('Quarter of the Church of the Joyful' and 'Quarter of the Ditch of Ghullān') are marked only with single-lined boxes, which may indicate an even lower level of defence, or merely an oversight on the part of the copyist. The three suburban quarters located nearest to the city are the 'Slav quarter' on the left, the 'Quarter of the Mosque of Ibn Siqlāb' and the 'Quarter of al-Tājī'. The labels for each of these quarters specifically mentions that they are surrounded by walls.

Beyond Palermo, Sicily is mere background. The mountains shown inland are only the peaks of the valley of Conca d'Oro near Palermo, and all the rivers and villages shown inland are also in the near environs of the capital city. It is only along the coast that we find labels for other Sicilian ports, such as Syracuse. The labels also indicate distances in miles between the different ports, which are shown in sequence but not in the correct position (this map is generally oriented to the north). Along the coast are a series of isolated peaks, with the volcanic Mount Etna shown on the lower left of the map, that is, to the south-west of the island, away from its correct position.

Who made this map? The author is not named anywhere in the treatise, but he repeatedly pledges allegiance to the caliphs of the Fatimid dynasty. The Fatimids were an Ismaili Shia dynasty, who established themselves on the coast of present-day Tunisia during the earlier part of the tenth century. They went on to take Egypt from their Abbasid Sunni arch-rivals in 969, where they founded their new capital city of Cairo in 971. By claiming descent from the Prophet's daughter Fāṭimah and his cousin ʿAlī, the Fatimid caliph imams saw themselves as the righteous leaders of all Muslims and guides to salvation. To fulfil the divine promise of salvation, the Fatimids relied on a clandestine missionary network, called *daʿwah*, which had been established in the mid-ninth century.

The training of missionaries became associated with the House of Knowledge (Dār al-ʿIlm), founded by the caliph al-Ḥākim in 1005, which became a focal point for encyclopedic learning by missionaries as well as non-Ismailis. As a result of caliphal patronage and the missionary network, Fatimid Cairo attracted some of the most influential scientists, philosophers and poets of the eleventh-century Islamic world.

An important clue for the dating of the treatise comes from the map of Sicily itself. Sicily is described and depicted as being under Muslim control, a bridgehead against the Italian mainland occupied by Islam's enemies. Therefore, the treatise must have been composed before the Norman invasion of 1068. That makes this map of Sicily the earliest map of the island to have survived to the present day from any tradition. The author also invokes curses at rebellious factions and tribal groups who sought to overthrow the Fatimids, both in North Africa and in Egypt. The dates of these rebellions allow us to pinpoint the composition of the work to between 1020 and 1050, some fifty to eighty years after the Fatimid conquest of Egypt.

The discovery of the treatise was so extraordinary that scholars needed to make sure that it was not a fake. World experts examined the manuscript in detail, searching for any industrial pigments or materials that would indicate a modern provenance. None was found. In fact, the paper type, the ink used and the handwriting were quite typical of manuscripts produced in Egypt and Syria in the twelfth and thirteenth centuries. This meant that the map was a copy made *c.*1200 of a treatise written by a Fatimid author some 150 years earlier, in the first half of the eleventh century. It has become evident that the manuscript is an invaluable testimony to an otherwise-lost medieval Islamic view of the world. After a long fundraising campaign to ensure that it would be accessible to the public, the Bodleian Library purchased the *Book of Curiosities of the Sciences and Marvels for the Eye* in 2002, and instigated a project of research and dissemination that continues to this day.

The Logic of Water

The structure of the *Book of Curiosities* departs from any prior Islamic or non-Islamic tradition, and shows a unique and ingenious approach to geographical material. First, the author enveloped his account of terrestrial geography with a cosmological introduction, choosing not to begin with the Earth at all. The first ten chapters of the treatise are devoted to the heavens and their influence on the Earth, accompanied by illustrative diagrams and drawings. From the outermost

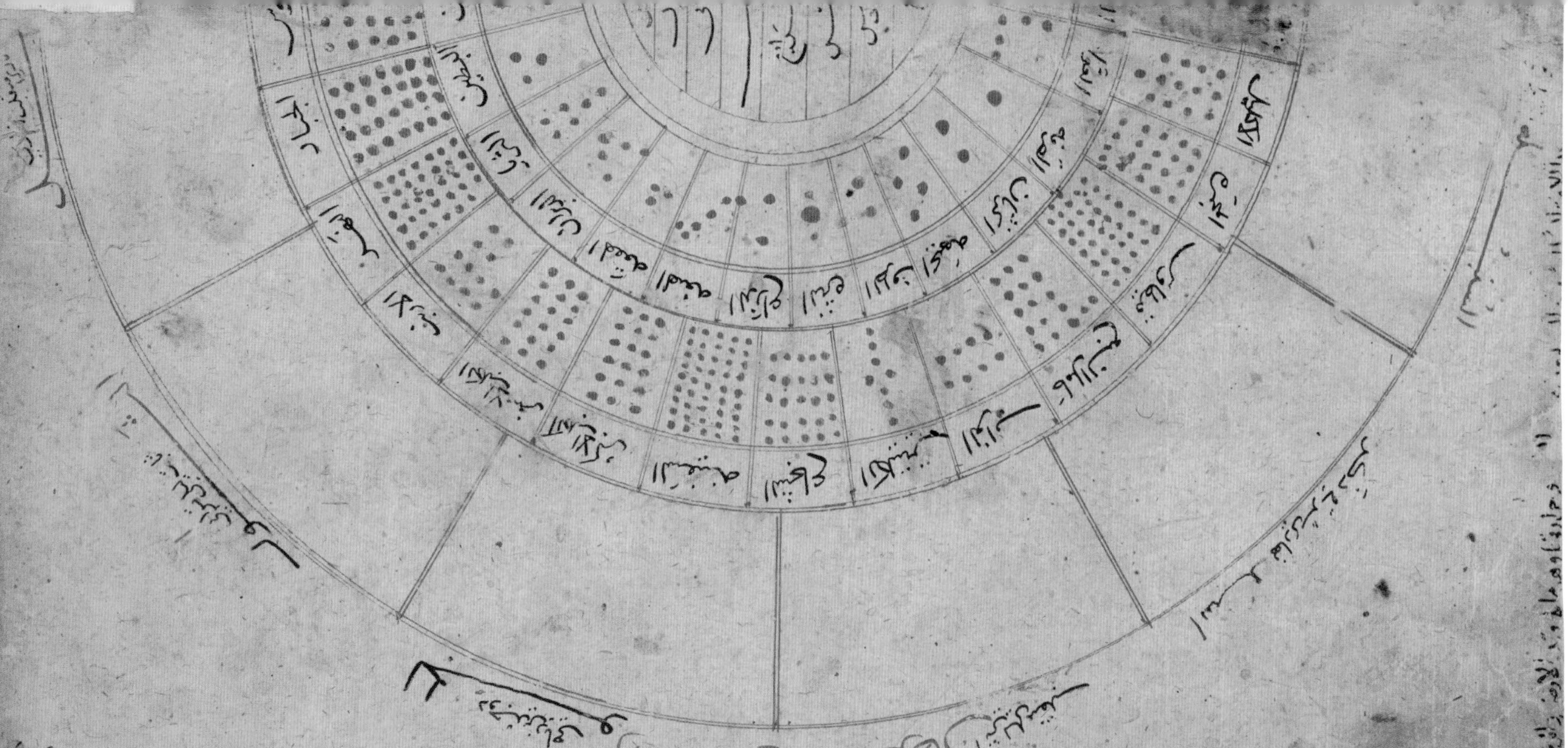

هذا الافق كثير اليبس لان الشمس اذا طلعت فيه نشفت البلة وحميت ويبست الرطوبة رطوبة الليل

واد هبتها والريح التي تهب من هذه الجهة تسمى الصبا وهي يابسة

قال الفرغاني في كتاب الفصول ان مساحة كل قطر من الفلك الاعظم الفي الف وماية الف وماية وتسعون ميلا

جميع ما في الصور الشمالية … عشرون كوكبا ثمانية واربعون كوكب والتي في الاقاليم الجنوبية وعلى تمام عشر صورة ثمانية احدى وعشرون كوكبا ومقسومة هذه الكواكب في العظم فمنها في الشرق

الاول … وفي الثاني مه وفي الثالث ج وفي الرابع لعد وفي الخامس ريز وفي السادس ط والباقية النجوم ه ومظلمه كا ع

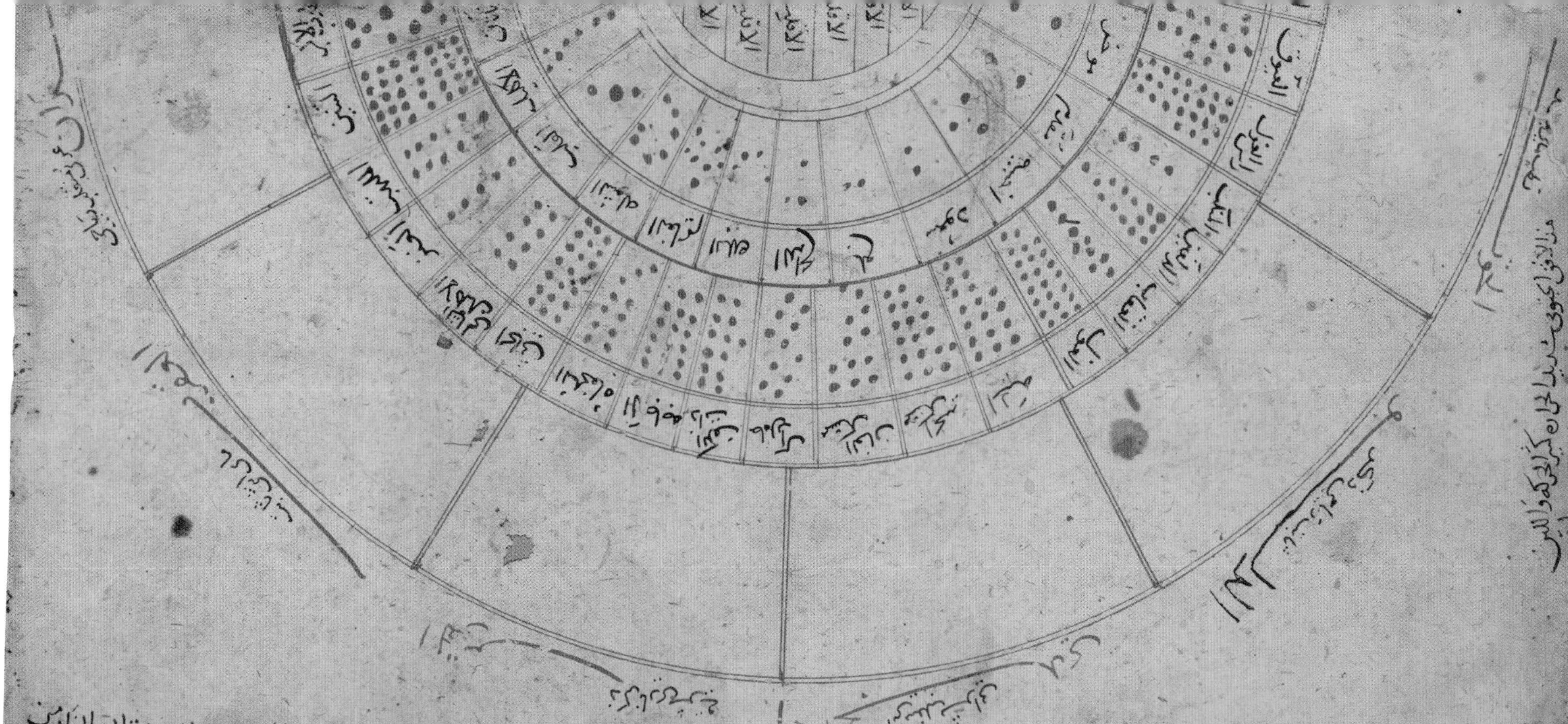

هذا الافق بطي لان الشمس اذا صارت اليه نزلت ما اجتمع من الندا والرطوبه التي صارت اليه بالنهار من
تسخن الشمس كان لها بالنهار وبددتها في الريح التي تهب من هذه الناحيه تسمى الدبور وهي رطبه ...

١٣٠٧٠٠٠٢٥

وذكر الفرغاني في كتاب الفصول في الفصل الواحد وعشرين ان قطر الفلك الاعظم مايه الف الف وثلثين الف الف وسبعمايه الف وخمسه عشر الف ميل واذا ضرب ذلك في ثلثه وسبع
كان دور الفلك الاعظم اربعمايه الف الف وعشرون الف الف وثمانمايه الف وثمانيه عشر الف وخمسمايه وسبعين ميلا
لجميع الكواكب التي قدر على معرفتها الف واثنين وعشرين كوكبا منها في صور البروج مايتان احد وستين كوكبا وفي الشماليه والجنوبيه ستمايه احد وستين ... وما في ... وسبعمايه ...
والخفيه وذوات الذوايب وارباب الحوادث والتغيرات والله اعلم بغيبه وما خلق وهو العزيز الحكيم

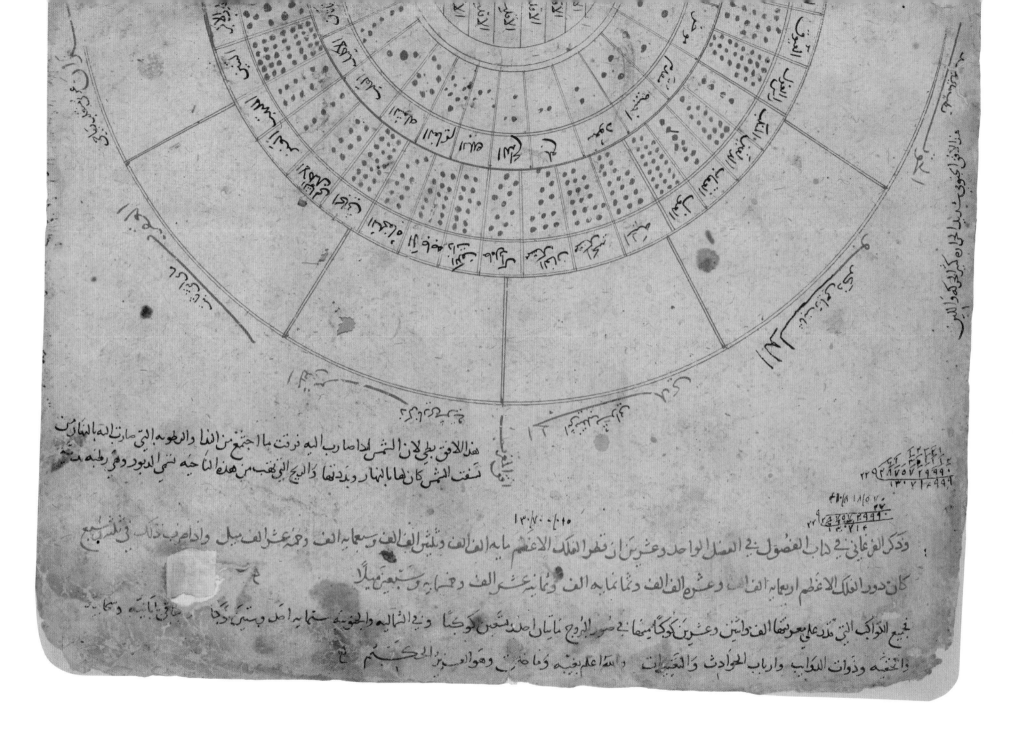

limit of the universe, the author methodically works his way downward through the various layers of celestial phenomena such as stars and comets until finally, in the last chapter, he reaches the Earth, where earthquakes and winds reflect the intersection of celestial and earthly events.

The first image in these chapters is a large circular diagram of the skies, occupying two facing pages. It is titled 'Illustration of the Encompassing Sphere and the Manner in which It Embraces All Existence, and Its Extent' (previous spread). This diagram reflects the prevailing geocentric conception of the universe inherited from antiquity. The Earth, with its seven climes, or zones of the inhabited world, is located at the centre of the diagram and at the centre of the universe. The twelve zodiacal signs around the outermost circle are named, in counterclockwise sequence, beginning with Aries at the top. The next innermost ring contains the names of thirty-six classical constellations – twenty-one northern constellations and fifteen southern ones – with each depicted as a pattern of dots representing stars. The subsequent inner two rings contain the names of the twenty-eight star groups known in pre-Islamic Arabian lore as the 'lunar mansions', because each of these star groups could be seen near the moon in one of its twenty-eight phases. The diagram is not a map of the skies as they appeared in the eleventh century, but an abstract map of the zodiacal signs and their approximate relationship to the classical Greek constellations, as well as of the smaller star groups traditionally used to tell the passage of time.

The second section of the treatise is devoted the Earth, and it is here that we find the most striking series of maps, including that of Sicily. The anonymous author opens this section on the Earth with a rectangular world map derived from late antiquity (discussed in Chapter 1). The author then provides maps of the three great seas known to him – the Mediterranean Sea, the Indian Ocean and the Caspian Sea. These are followed by maps of islands and peninsulas (all in the Mediterranean), by maps of bays (in the Aegean) and by maps of some dozen different saltwater and freshwater lakes of the world. Finally, the last maps are of the five great rivers known to the author: the Nile, the Euphrates, the Tigris, the Oxus and the Indus.

In the *Book of Curiosities*, water and waterways are the organizing principles. While al-Iṣṭakhrī presented us with interconnected regional maps of the Muslim world, the anonymous Fatimid author sees the world as a sequence of overflowing seas, lakes and rivers. Fundamentally, the logic of the *Book of Curiosities* is a logic of water. The waterways dictate and bound the focus of each of the maps. If

(previous spread) A circular diagram of the skies, entitled 'The Illustration of the Encompassing Sphere and the Manner in which It Embraces All Existence, and its Extent', from the *Book of Curiosities*, copied *c.*1200. Bodleian Library, University of Oxford, MS. Arab c. 90, fols 2b–3a.

(following spread) Map of the Mediterranean Sea from the *Book of Curiosities*, copied *c.*1200. Bodleian Library, University of Oxford, MS. Arab c. 90, fols 30b–31a.

al-Iṣṭakhrī's maps are devoted, literally, to the lands of Islam, in the maps of the *Book of Curiosities* we are never far from the sea.

This emphasis on the sea is very striking in the map of the Mediterranean (pp. 74–5), which is presented as a perfect oval. There is no attempt to trace any of the actual coastlines, not even the Iberian Peninsula. Like al-Iṣṭakhrī's representation of the Mediterranean, or perhaps even more so, the form is completely abstract. But, unlike al-Iṣṭakhrī, the Fatimid author's primary interest is the sea, not the land. The map of the Mediterranean in the *Book of Curiosities* is a depiction of maritime space. Its focus is the coast to the exclusion of any inland features. There is nothing else in this Mediterranean but islands and mooring points: there are 118 islands within the dark green sea and 121 harbours and anchorages on the rim. The oval seems to be oriented to the north. The Strait of Gibraltar, indicated by a thin red line, is at the far left of the oval. The harbours and anchorages of western Europe and Anatolia are in the upper half, while those of Palestine, Egypt and North Africa are at the bottom.

The extreme abstraction of the coastline notwithstanding, this is in fact a mariner's view of the Mediterranean. The label for each harbour contains unique information on capacity, defensive installations and the availability of fresh water. Protection from prevailing winds is mentioned for a large number of the harbours and anchorages, with Greek nomenclature used for wind navigation. Some of the best anchorages, such as Tyre and Caesarea, are said to offer protection from all winds. There are also sailing distances, given for travel in the open seas, measured in days and nights of sailing. As for the islands in the interior of the oval, their position bears no relationship to the sequence of harbours on the rim. Almost all the islands in the middle of the map are merely mentioned by name and represented as small circles of equal size. Only Sicily and Cyprus, represented as rectangles, have longer descriptive labels, which correspond to some of the material on the individual maps of these two islands in the following folios (p. 76).

The *Book of Curiosities*' maritime logic is directly related to the political interests and orientation of the Fatimid caliphate as a Mediterranean power. More than previous Islamic empires, the Fatimids relied on naval power and maritime commerce to pursue their political and religious ambitions. Their conquest of Egypt in 969 was achieved through a combined naval and land attack, with galleys entering the eastern and western arms of the Nile Delta through the ports of Tinnīs and Damietta, then sailing down the Nile. A major naval campaign against Byzantium set out in 996, and a squadron of galleys suppressed a rebellion in

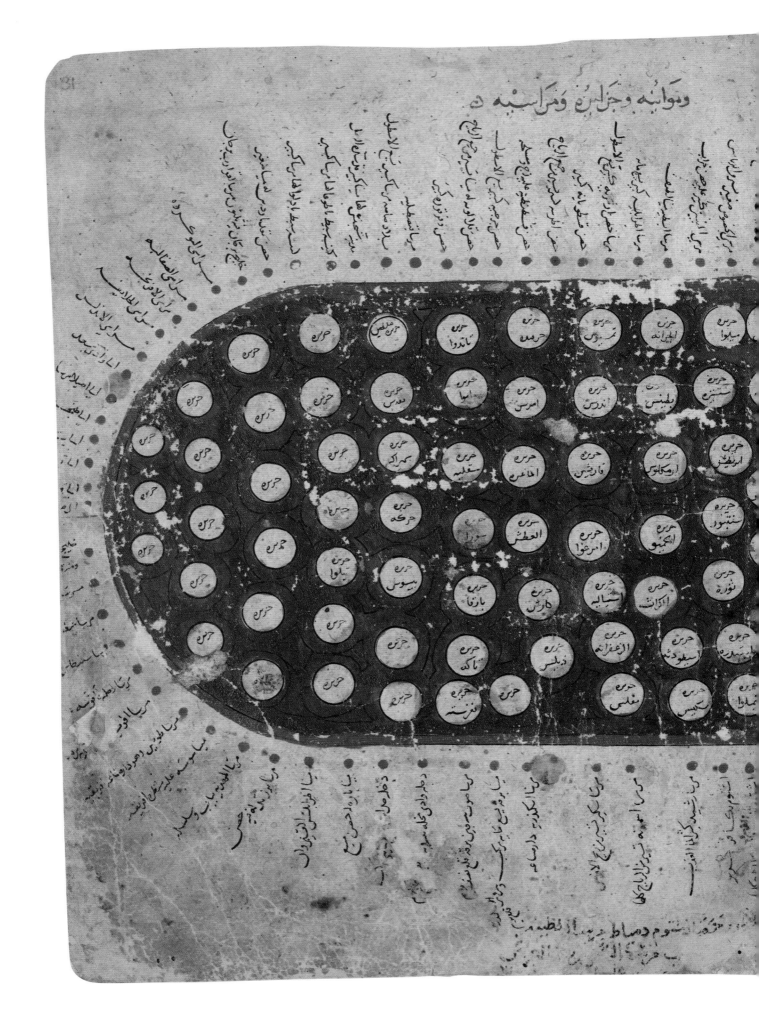

وموانيه وجزايره ومراسيه

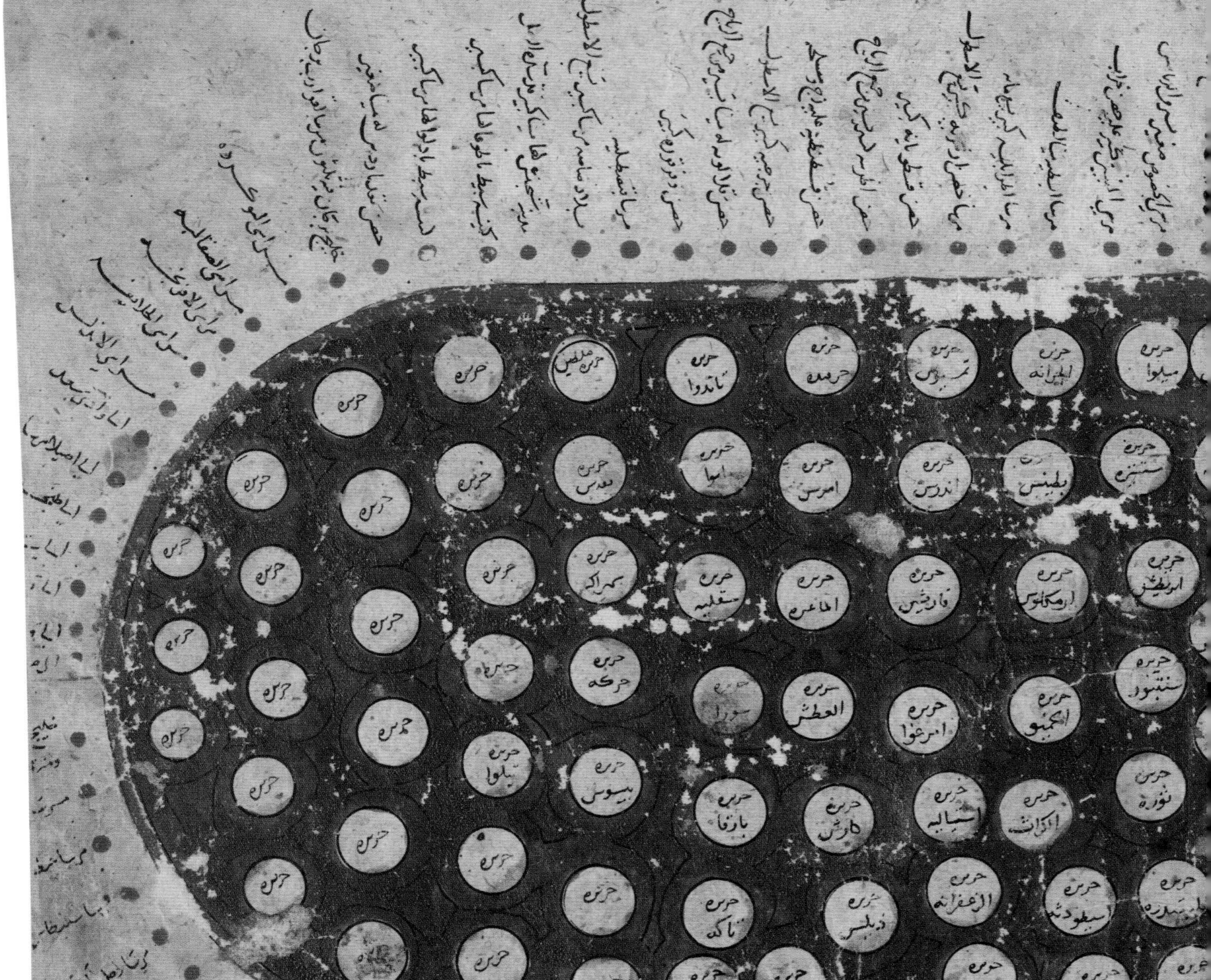

الفصل العاشر في البحر الغربي وهو الشامي

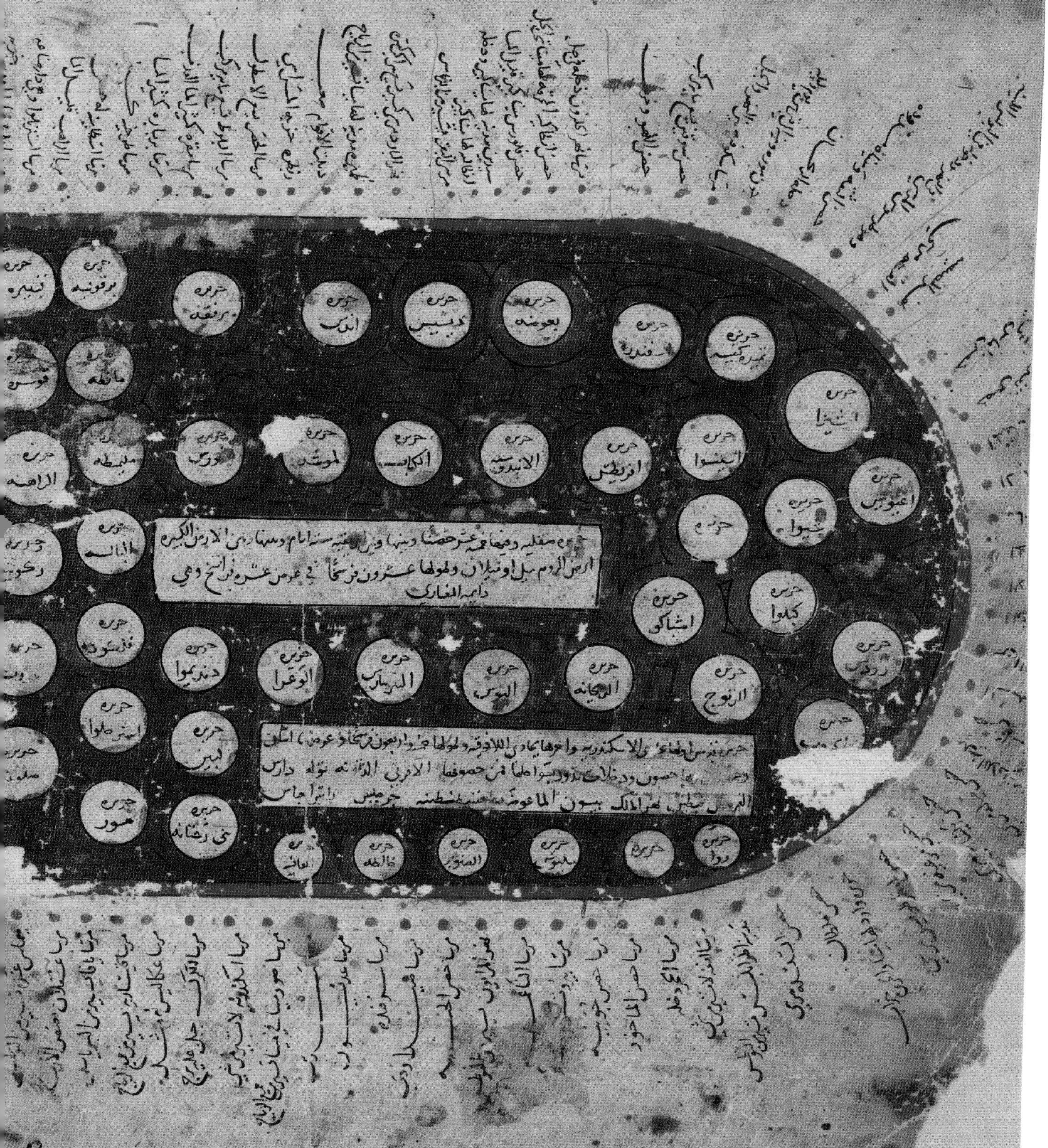

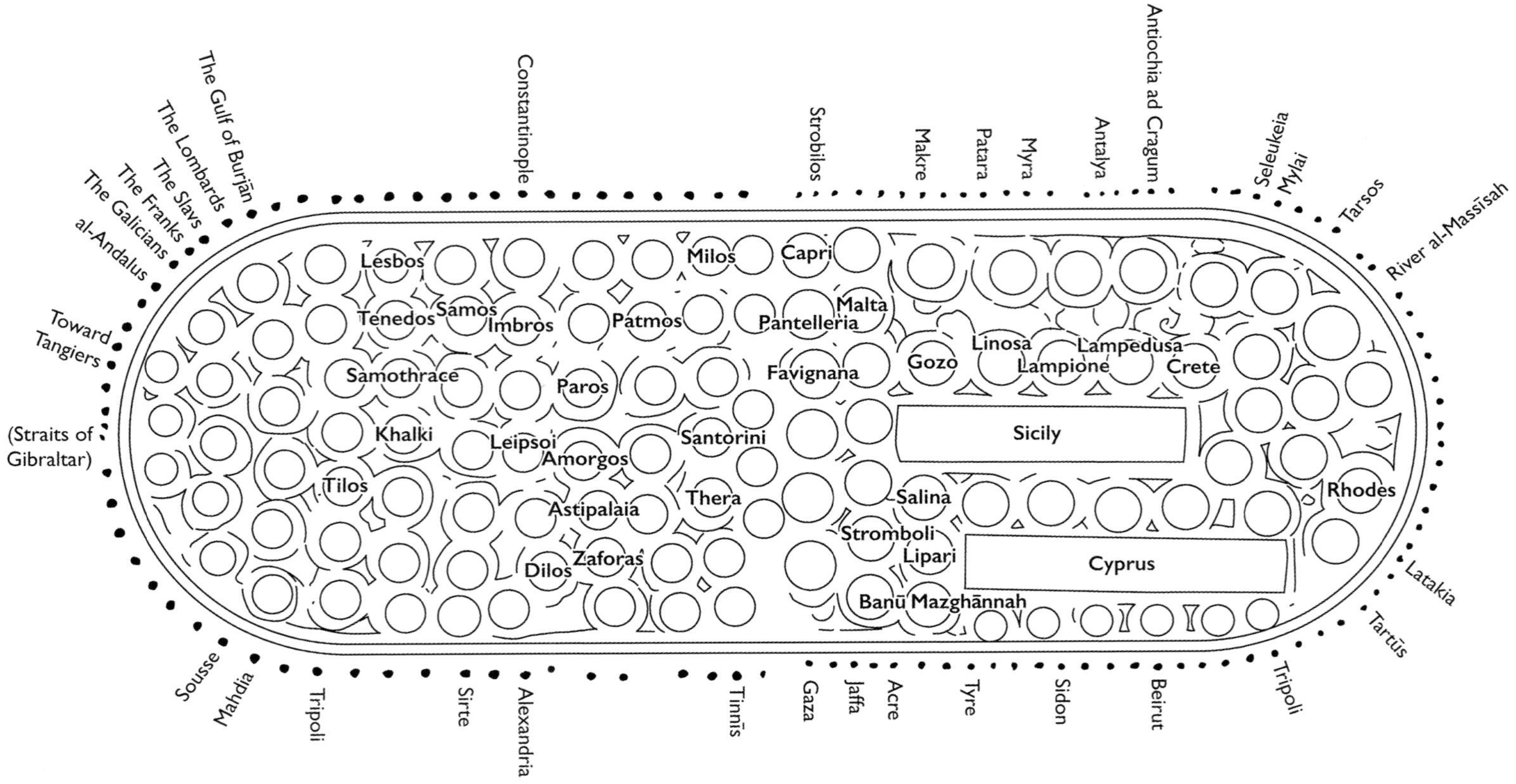

Explanatory diagram of the Map of the Mediterranean in the *Book of Curiosities*, copied *c.*1200.

Acre in 998. The regularity with which the Fatimids sent navies to the Syro-Palestinian coast, despite occasional disasters and the scarcity of wood, shows their determination to have a military presence in the Mediterranean.

One of their prime targets was the strategic island of Cyprus. The *Book of Curiosities* contains a remarkable diagram of Cyprus, which is the earliest known detailed map of the island from any cartographic tradition (p. 78). This map of Cyprus belongs to a chapter on the 'Islands of the Infidels', devoted to Mediterranean and Indian Ocean islands in the hands of non-Muslims. These included Cyprus, which used to be shared between Muslim and Christian forces until it fell under full Byzantine control in 965. The island is represented by a square surrounded on all four sides by a strip of green-coloured sea. Twenty-five cells along the edges, or 'coasts', are meant to represent anchorages or harbours, although some have been left empty. Nine additional anchorages or harbours are represented by a strip of nine cells in the middle of the square, the strip entitled 'Names of the remainder of its harbours'. There seems little doubt that the map-maker was forced to place these harbours in the middle of the square as a result of the lack of space along the edges.

In spite of the strange design, the harbours of Cyprus are listed in roughly the correct sequence. The harbours on the northern coast are represented on the middle strip, starting from the north-western Acamas Peninsula on the left-hand side of the strip. Acamas is described as the 'Beginning of the island'. The sequence proceeds along the strip until it reaches the anchorage of Akraia, located at Cape Apostolos Andreas in the north-eastern tip of Cyprus. The list then proceeds in counterclockwise order along the four edges of the rectangle, representing harbours on the eastern, southern, and western coasts of the island. It culminates with Paphos on the eastern coast, at the bottom right.

The type of information provided for each of the harbours and anchorages of Cyprus is very similar to that provided in the map of the Mediterranean, though it is often more expansive. In particular, protection from the prevailing winds is systematically recorded. The winds named are the Boreas (north), Notos (south) and Euros (east or south-east), as well as a wind called the Frankish, *al-Ifranjī*. Capacity, water sources and prominent landmarks are again frequently mentioned. The ports of Paphos and of Jurjis (probably Hagios Georgios, a monastery east of present-day Limassol on the southern coast) are said to accommodate 150 vessels each. Sailing distances are given here only for sailing on the high seas, with no distances provided between adjacent anchorages or harbours. Sailing from Akraia in Cyprus to Rhodes is said to take one day and one night, but required propulsion by the Boreas (northerly wind). The author was clearly aware of the wind patterns in the Mediterranean; sailing away from the coast was much more difficult when travelling in a counterclockwise direction, for example from Egypt to Anatolia.

The maps of the Mediterranean and of Cyprus are obviously abstractions, diagrams more than maps, where the coastlines and even the major gulfs are not represented at all. At the same time, anchorages and harbours are listed in geographical sequence. The key data for each anchorage or harbour are the protection from the four prevailing winds, indications of their size and quality and the availability of fresh water. These maps of the Mediterranean and Cyprus are drawn from the perspective of the sea. This does not mean that the maps as we have them were meant to be taken on board a ship. But, in all probability, the map-maker was collecting information from records kept by sailors. He also made a conscious choice to depict only information that was relevant for seafarers and to opt for absolute abstraction.

These radical choices of both form and content are interrelated. The total abstraction of the coastline – the perfect oval for the Mediterranean – is chosen precisely because the maps were intended to simplify the presentation of maritime space. The omission of all gulfs was not done out of a landlubber's ignorance. On

... الخامس عشر في جزائر ...

ليس الغرض في كتابنا هذا صفات سائر الجزائر المسكونة وانما القصد فيه لمع من ذكر كل فن مما غرب فهمه وقد اوردنا في كتابنا المسمى بالمحيط صفات جزائر البحار كلها حسب الطاقة والى ما انتهى اليه علمنا والله المعين على رضا المصنف بمنه وكرمه

صورة جزيرة قبرص ومراسيها

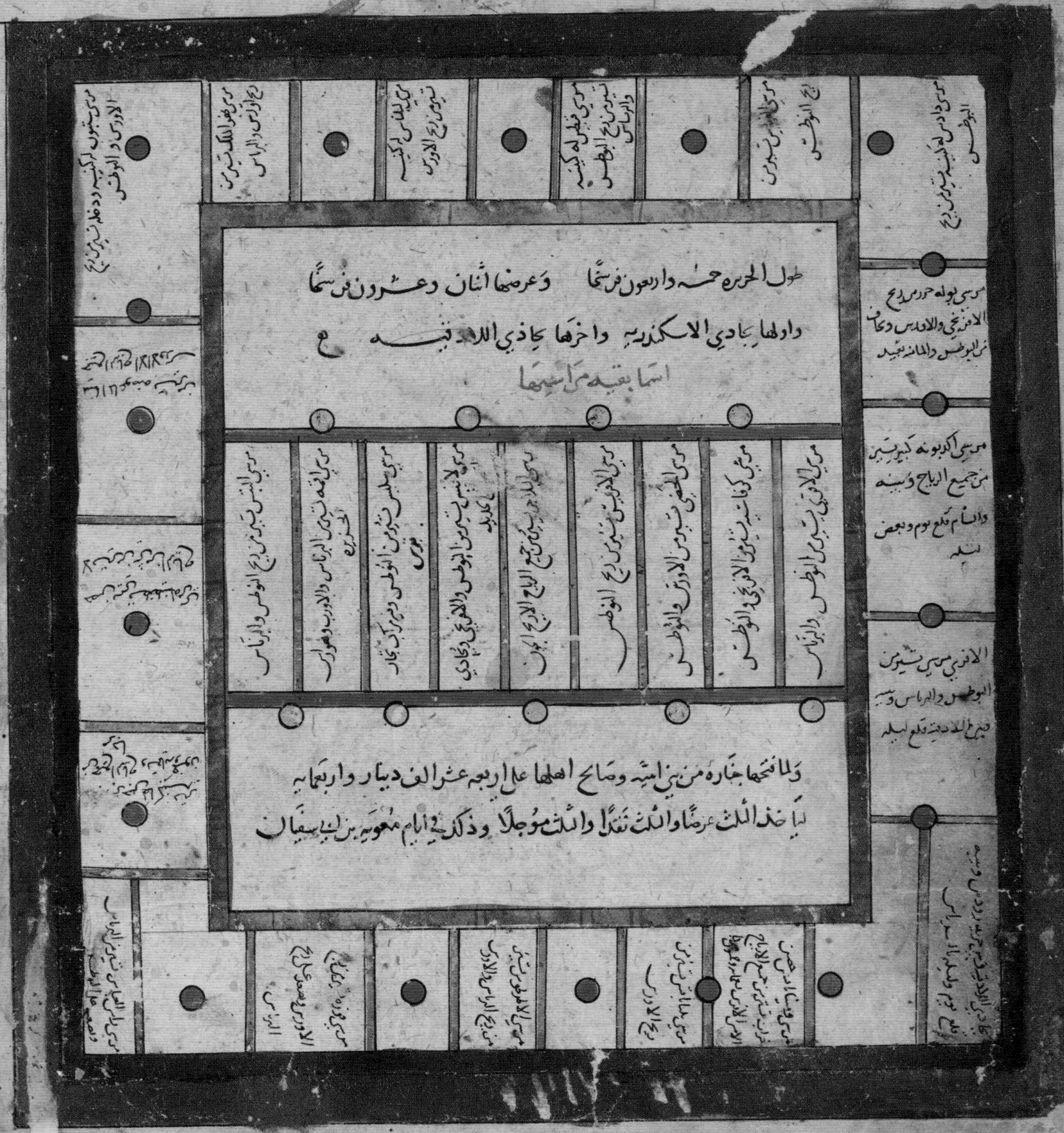

ومن الجزيرة المصطكي واللاذن والشب اليابسه والطرية والزاج والقلقنت والقلقديس وما يجلب من ارض الروم الى كل بلد من سائر المدن وغيرها

Map of the island of Cyprus from the *Book of Curiosities*, copied *c.*1200. Bodleian Library, University of Oxford, MS. Arab c. 90, fol. 36b.

the contrary, these maps carry much information that is relevant to navigation, more than any other map known to us before the age of the Crusades. But, when the maritime journey was by default a voyage along the coast, the key information is the sequence of harbours, their quality and any recognizable markers that would have been visible from on board a vessel. The distances between adjacent harbours are not indicated, perhaps because they were usually short. However, sailing distance is mentioned for travel in the high seas between ports that are not next to each other. Here the map surpasses a simple textual narrative of a maritime itinerary because it creates connections between ports that are not located next to each other. A map, unlike a text, can allow the viewer to read the navigation material in ways that are not only sequential or linear.

Representations of Fatimid Power

The maps of the *Book of Curiosities* are not just about the sea but also about power. This is very apparent in one of the most visually striking maps from the *Book of Curiosities*, the map of the port town of Mahdia on the North African coast (p. 81). Mahdia is today no more than a small resort town in Tunisia, but in the eleventh century it was nearly as important as Palermo. Its artificial harbour was bustling with ships sailing westward from Alexandria, and its significance for the Fatimid empire was not only strategic but historical too. Mahdia was the first capital built by the Fatimid caliphs when they were still a North African empire, and the city retained its symbolic importance for the dynasty even after its move to Egypt and the foundation of Cairo. Fitting with this legacy, the map of Mahdia in the *Book of Curiosities* is an iconic image of Fatimid power. A millennium later, it still has a distinctive charm. The Bodleian Library has elected to reproduce it on the mugs and scarves it sells, and it was also chosen as the cover for an authoritative history of the Ismaili sect.[2]

In the map of Mahdia, as in the map of Sicily, the focus is on the pictorial representation of three man-made complexes: the walls, the harbour and the twin palaces. The map is oriented to the east, to best fit the elongated peninsula into the page. Rectangular stone walls surround the city, pierced by two large gates in the west (at the bottom of the map). An indentation in the northern walls follows the contours of the coast. An arch in the south-eastern corner of the map, topped by another structure or structures, is labelled the 'Harbour'. The two isolated and elaborate buildings near the southern walls of the city are labelled 'Palaces of the imams, may they rest in peace'.

Map of the port city of Mahdia, in present-day Tunisia, from the *Book of Curiosities*, copied *c.*1200. Bodleian Library, University of Oxford, MS. Arab c. 90, fol. 34a.

The general layout of the city in the map corresponds to what is known about the actual layout of eleventh-century Mahdia. Our map-maker gives the impression that the map of Mahdia was drawn by someone looking at it from the outside, perhaps by a person on board a ship making its way to the city harbour. The indentation in the northern walls is clearly an attempt to represent the curve of the northern coast of the peninsula. The location of the harbour in the map also corresponds to the actual location of the artificial harbour, dug in the bedrock of the south-eastern corner of the city. The visual representation of the harbour tallies with European reports on the existence of an arch above the entrance, linking two towers that guarded the passageway. The entrance, which can still be seen today, forms a passage about 15 metres wide.

The twin palaces of the Fatimid imams are also correctly located near the southern walls of the city, and are made to look as if they are facing each other. The palace on the right has its gate facing south, while the palace on the left is depicted with a large rectangular blind niche, suggesting that it is represented from the side or from the back. Such a presentation corresponds to contemporary descriptions of the twin Fatimid palaces that dominated the landscape of the city. Although archaeological excavations found very few remains of these buildings, it seems clear that the map-maker attempted to capture the position of the two palaces in relation to each other. It is also notable that some attention has been given to individuating each of the structures: the one on the right consists of a tower surrounded by high walls, while the one on the left shows a wider central building.

The imposing twin gates at the western walls were noted by most medieval travellers to the city. Ibn Ḥawqal, visiting Mahdia during the second half of the tenth century, stated that the city walls had two doors 'the likes of which I have not seen on the face of the Earth', except in Raqqa in Syria.[3] Al-Idrīsī (d. 1165) lauds the two iron gates as two of the world's wonders.[4] The depiction of the two gates on the map is evidently not incidental or ornamental, for it fits neatly with the impression it made on medieval visitors to the city. Like the depiction of the port and of the palaces, this image aims to capture the distinctive features of the actual walls.

The list on the top left, written in the open space within the walls, is a maritime itinerary between Mahdia and Palermo. It consists of the names of anchorages along the way and the distance between them in miles. The first thirteen stops are along the coasts of Ifrīqiyah, present-day Tunis, up to Kelibia at the tip of the North African mainland and the island of Pantelleria, and from

من بعده البرهان والحجج   جاؤوا وقد رمدت أبصارهم حنقا وقد تحلت بكحل الرهج ما ناحفهن مع الرجال حتفهم

على مقدمة الغوغا والهمج   لا تفرحن ضيق المسلك منفرجا فان صبح الهدى يلمع لي الا فرج ۞

كم مرة ترقى درجا لم تدر حتى بها زلت به رجله من اوسط الدرج ۞ ثم كان من امره ما قدره الله تعالى من امره وسيره و[illegible] على يد المنصور بالله امير المؤمنين

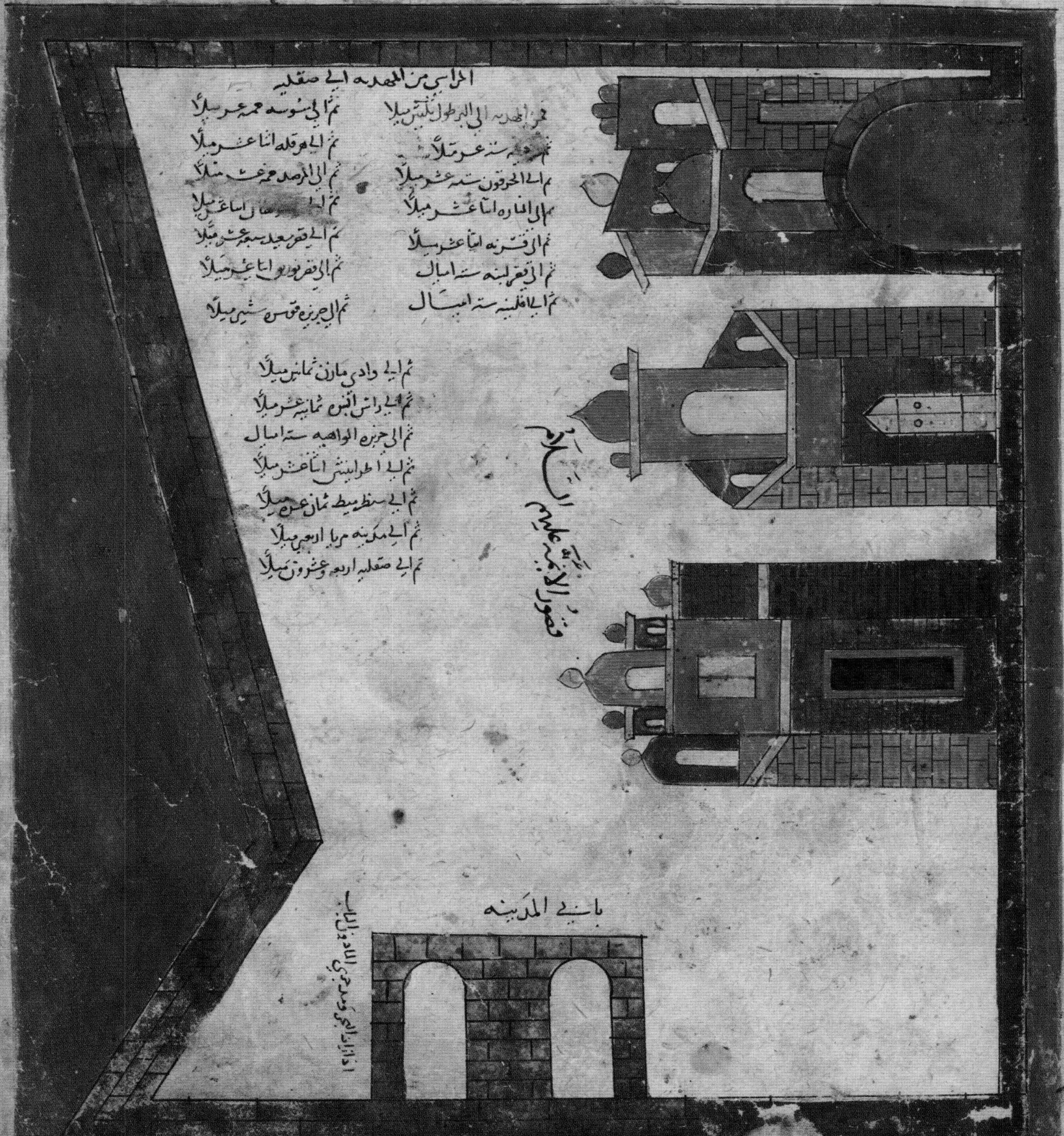

...ب هذه المدينه تنيس بيت صابن بدارس اخر ملوك القبط وكانت بحيره اجنه وخليج تخترقها من ما النيل
...ضياع عامره وزروع متوازنه الى ان غلب عليها البحر وقد هاج فهجم من فم الاشتوم على اراضيها
...عمايرها تغرقها فما كان من ارضها مستفل هلك وعلاه البحر وما كان منها على كوم عالي مثل تنيس
...ونه وغيرها فما هو باقي لم يعلوه الما وبقي على اسمه وكان ذلك الغرق قبل الاسلام بمايه سنه وقد ذكر
...سعودي في كتابه مروج الذهب بنفد التجار الى الفقار وقد شاهدنا ذلك في عصرنا فمن ذلك
...ل على صحه قوله وما استجر في طريق الجفار من مواضع كانت تعمر فصارت بحرا وذلك
...تقدير العزيز العليم

في هذا الربض مساجد وكنايس ومغارش لسبيض الامتعه وحجاره مصرسه منقوشه
لقرب الباب وفقايها وفيها هدف للرماه
وطالع تاسيس هذه المدينه برج الحوت وصاحبه المشتري السعد الاعظم وصاحبه الشرف الزهره
ولذلك كثر طرب نفوس اهلها وفرحهم ورغبتهم في مداومات اللذات واستماع الاغاني ومواصله
المسرات والرغبه في الراحه والمراح مايجب التعب والمشقه واحب للنقش والصور والرقم والتلوين
بالاصباغ وعلى قلة الفحص في السعر وترك المخالفه لمن يصاحبون وكثرة المباهات بالفنون
وحسن الموازره لمن يستخدمهم ومحبتهم للغربا والمسافرين والمواطنه على معاشرتهم وسرور
بمكاسبهم ونفعهم وتركهم الحسد لمن يجوه والعتب على زلته ويمدحونه ويفضلو
ويلوسون انفسهم في التقصير عن واجباته وما يستحقه والقيام بذلك ع
في هذا الربض دواليب تنقل الماء الى الصانع واحمامات وديوان كبير للسمك ع

مدن البحر

there to Mazara del Vallo on the south-western coast of Sicily and onward to Palermo. The function of this maritime itinerary, located as it is in the blank intramural space of Mahdia, is to underline the strategic location of this port city.

There are strong visual and thematic links between the map of Mahdia and the map of Sicily/Palermo. Mahdia, the first Fatimid capital, served as the model for Fatimid building projects in Palermo. The same caliph, al-Qāʾim bi-Amr Allāh, who ordered the building of the palatial al-Khāliṣah compound in Palermo also constructed one of the twin palaces shown on the Mahdia map. More generally, both were seats of Fatimid government near the major ports of the central Mediterranean. In both cities, palatial areas were surrounded by larger suburban areas where commercial activities took place. While al-Khāliṣah abutted the old city of Palermo, palatial Mahdia depended on the unwalled residential town of Zawīlah, which is not shown on this map but is mentioned in the preceding text.

We can now understand why the map-maker chose to represent Sicily the way he did: his map is a cartographical manifestation of a political statement about Fatimid ambitions in the Mediterranean. It aimed to convey the impregnability of the fortifications of the island and its capital Palermo, and to demonstrate the strength of the Fatimid hold over the island. Its visual centre is the seat of power in Palermo and its suburbs. The map highlights the fortifications of a port city: gated walls, a well-protected palace and chain towers at the entrance to the harbour. These three key elements are represented prominently and pictorially. Sicily, the map-maker tells us, is not for the taking.

The author of the *Book of Curiosities* included a third image of a Mediterranean island-city, a diagram of the lost city of Tinnīs in the Nile Delta (previous spread). Tinnīs was located on an island in a brackish lake formed by the meeting of the Mediterranean and an eastern arm of the Nile Delta, near present-day Port Said. In its heyday it was a major centre for textile production, as well as the major port on the north-eastern branch of the delta, and a natural stopping point for trade between Egypt and the Syro-Palestinian coast. It was repeatedly attacked by Crusader fleets until Saladin ordered it to be evacuated from its civilian population in 1192–3. The city was eventually completely abandoned during the course of the thirteenth century, and its remains are now entirely covered by sand. The map of Tinnīs in the *Book of Curiosities* is therefore like a map of a sunken Pompeii, a gift from heaven for archaeologists in the early stages of trying to reconstruct the site.

(opposite) Map of the island-city of Tinnīs from al-Qazwīnī's *Monuments of Nations*, copied 1329.

(previous spread) Map of the island-city of Tinnīs, in the Nile Delta, from the *Book of Curiosities*, copied *c.*1200. Bodleian Library, University of Oxford, MS. Arab c. 90, fols 35b–36a.

جاشك جزيرة اهله بقرب جزيرة تنيس برهوت بارض حضرموت وقد مر ذكرها في حضرموت

لاهلها جلادة وجرءة في حرب البحر وعلاج السفن جلاده ليس لغيرهم مثلها حتى ان الواحد منهم يسبح في

الماء اياما ويجالد بالسيف مجالدة من هو على الارض ويقول اهل تنيس ان بعض ملوك الهند اهدى

الى بعض الملوك جواري فلما وصل المركب الى جاشك خرج الجواري سفينة فاحتطفهن الجن وافترشوهن

فولدن الذين بها فلهذا ياتون بما عجز عنه غيرهم **جاطة** جزيرة على مرحلة طرفة من ارض

افرنجة طولها مائة اميال وعرضها خمسة اميال بها ثلثة اعين عذبة الماء وبها مزارع واثار قديمة وبها من

الايل ما لا يحصى حدثني الفقيه سليمان الملياني رحمه الله ان بها عنزا كثيرا انسية توحشت اذا قصدها قاصد

اهوت نفسها من جبال شاهقة وتقع على قوائمها بخلاف الايل فانها تقع على قرونها **جزيرة**

تنيس جزيرة قريبة من البرين فرما ودمياط في وسط بحيرة مفردة عن البحر الاعظم بينها وبين

البحر الاعظم بر مستطيل وهي جزيرة من البحرين واول هذا البر قرب الفرما وهناك فوهة يدخل منها ماء

البحر الاعظم الى بحر تنيس في موضع يقال له القرباج وهو يحول بين البحر الاعظم وبحر تنيس سائر في ذلك البر

ثلثة ايام الى قرب دمياط وهناك فوهة اخرى ياخذ الماء من البحر الاعظم الى بحيرة تنيس وبقرب ذلك

الفوهة النيل يصب الى بحيرة تنيس والبحيرة مقدار اقلاع يوم في عرض نصف يوم ويكون ماؤها اكثر السنة ملحا

لدخول ماء البحر اليه عند هبوب الشمال فاذا انصرف نيل مصر عند دخول الشتاء وهبوب الرياح الغربية

دخلت البحيرة وحلا سيف البحر الملح مقدار بريدين وعند ذلك تكاملت النيل وغلبت حلاوته ماء البحيرة

فصارت البحيرة حلوا فحينئذ مدخر اهل تنيس المياه في صهاريجهم ومصانعهم لشرب تمام السنة وهذه صورتها

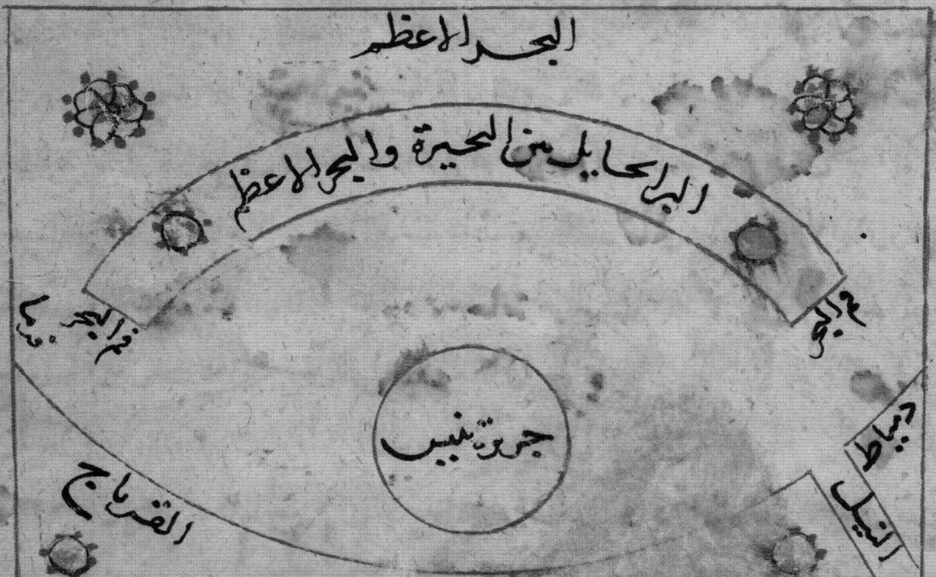

ذكروا انه ليس بجزيرة تنيس شيء من الهوام المؤذية لان ارضها سبخة شديدة الملوحة وقد صنف في اخبار

تنيس كتاب ذكر فيه انها بنيت في سنة ثلثين ومائتين بطالع الحوت اثنا عشر درجة حد الزهرة وشرفها

This map of the town of Tinnīs pictorially represents only the walls and the ports of the city. The diagram extends over two folios and shows the city with the green Mediterranean at the top of the map and, on the other three sides, the blue deltaic lake, called Lake Manzala, that encircled the island-city. The rectangular walls of the city have twelve gates, which are indicated by simple squares, without doors or arches. Each of the western and eastern sides has two gates, and four more gates are located in the northern side. The southern walls have six gates, two of which regulate the water channels that flow into the city. The labels near these twin gates indicate that this is the entry to the harbour. The one on the left reads 'This is a harbour for ships, on which there is a gate', while the label on the right reads, 'This is a harbour into which ships enter.'

The labels around the walls of Tinnīs refer to suburban areas: the governor's palace, the prayer grounds for religious festivals and the fishermen's shacks near the harbour inlets. As in the map of Mahdia, however, the intramural area is devoid of any topographical detail or iconography. Instead, the vast space inside the city walls is used for a text devoted to the favourable traits of the city's inhabitants and its pre-Islamic history: the people of Tinnīs, we are told, are full of happiness and joy because the city was founded when Pisces was in the ascendant. In the preceding folios of the treatise the author's text describes markets and textile workshops, mosques and inns, but none of these are indicated on the map.

The map of Tinnīs in the *Book of Curiosities* emphasizes walls and ports. Its distinctive political message comes into sharper relief when we compare it to another map of the same island (p. 85), included in the *Monuments of Nations* by the Iranian author al-Qazwīnī (d. 1283). Al-Qazwīnī's map, made after the city had been abandoned, depicts the opening of Lake Manzala towards the Nile at the bottom right and two openings towards the Mediterranean on both sides. The stretch of land that separates Lake Manzala from the Mediterranean is depicted as a semicircle at the top of the diagram, and the island of Tinnīs itself as a circle within the lake. Because the map-maker in al-Qazwīnī's treatise is interested in the unusual location of Tinnīs on an island within a deltaic lake, the walls of the city and its port are not mentioned or represented.

The maps of Sicily/Palermo, Mahdia and Tinnīs form together the earliest set of town plans to have survived from medieval Islam. There are not many other urban maps from the Middle Ages. The most striking example of one is a plan of another peninsular port town, in this case Aden on the south coast of Arabia, one of the most important commercial entrepôts in the Indian Ocean (pp. 88–9). This plan was made around 1220 by a traveller and merchant called Ibn al-Mujāwir, who has left us a most engaging geographical and historical account of the Arabian Peninsula. Unusually, Ibn al-Mujāwir's text is accompanied by a set of thirteen schematic town plans or maps, depicting, among other towns, Mecca, Zabīd and Taᶜizz, mostly using squares and circles, with minimal detail and little attempt to capture architectural features.[5] This set comes from a peculiar and understudied local map-making tradition that emerged in the Arabian Peninsula in the Late Middle Ages. A fiscal text commissioned for a fourteenth-century Yemeni sultan, which has been recently published, is accompanied by a set of maps of administrative regions and important port cities, executed in a style that bears close resemblance to that of Ibn al-Mujāwir.[6]

The map of Aden reproduced here is the most complex in Ibn al-Mujāwir's treatise, and clearly foregrounds the commercial pursuits of the author and map-maker. The circle at the bottom of the map is labelled 'Balance', and is probably the weighing house at the entrance to the port. The label prominently placed above the straight horizontal line that runs across the diagram reads 'Custom house'. Aden was a necessary stopping point in the Indian Ocean trade, and for a merchant coming to Aden the custom house was the most visible – and dreaded – urban institution, for it was where one's cargo was searched and where goods could be seized. The long label on the left describes another feature closely connected to international trade: a south-facing watchtower from where the Egyptian convoys of merchant ships could be spotted.

Ibn al-Mujāwir tells us that he drew this view of Aden looking west from the fortress of Sīrah, an islet guarding the entrance to the main port of the town. The peninsula on which Aden lies is drawn as a perfect circle, with the Indian Ocean to the south and the bay known as the Lake of the Persians in the north-west. As in the map of Mahdia, the port town is made to look like an island, surrounded by water on all sides, and the message in both cases is that of a well-protected domain. Aden itself is represented by a label in the centre of the map, but not pictorially, and there are no walls or fortifications, although the different diagonal rectangles that dominate the diagram represent the steep mountains that overlook Aden on all sides.

Perhaps the most striking aspect of the urban plans from medieval Islam is the emptiness at their centre. The representations of Palermo, Mahdia, Aden and Tinnīs all show the boundaries of the town, the fortifications or the custom

قديمه طولها اربعون ذراعا وبير زعفران اشتريت بمدته واوقفت على
المسلمين ٥ فصل حدثني عبد الله بن محمد بن يحيى قال انه كان ينقل
ما بير زعفران الى بلاد اليمن قال لان سيف الدين اتابك سنقر مولا الملك
المعز اسمعيل بن طغتكين شرب عند المعتمد محمد بن علي التكريتي نبيذا اعجبه
طعمه فقال له مم عملت هذا النبيذ قال من ما زعفران اذا قلت في هذا الماداد
وترك في الشمس يرجع نبيذا كما ولا يحتاج الى عسل ولا الى شي اي وضعه فمن الحين
كان ينقل له هذا الما الى الجند ونحو صنعا وزبيد يعملون منه نبيذا
والاصح ما الزب ويقال انه في الاصل كان عذبا فراتا والان قد علته ملوحة
بعض الشي من سو افعال الخلق وبير السلامي بير حفرها الشيخ اسمعيل بن عبد الرحمن
السلامي وبير روح قديمه وبير عود قديمه وبير بن الدوس صهر الشيخ معمر
بن جريح وبير الحمام حفرها محمد بن علي التكريتي وبير الحمام الثانية قديمه
وبير مور قديمه وبير جلاد قديمه وبير الحصامي قديمة فصل حدثني
محمد بن ركل بن الحسن الكرماني عن رجل من اهل عدن قال حدثني عبد الله
بن محمد الاسحاقي الداعي ان بداخل عدن مايه وثمانون بيرا حلوة ولكنها
مانعه والله اعلم ذكر الابار المالحة بعدن بير وصاح
قديمه وبير بانيه الى جنبها وبير بن عند مرابط الجمال وبير ام حسن
قديمه وبير قند له على طريق الباب وبير سنبل قرب الحمام وبير سالم وبير
حندود وبير فرج وبير الزنوج وبير الافيله وحفرت سنة عشرين
وستماية وبير دلش السواحي وبير في قرب دار القطيعي السلاطه وبير
الشريعة ذكر ابار ماوها بخر عدن بير في حافة الدبا كله
وبير عند باب مكسور وثلثة ابار للبرابر وبير عند الجامع وبير

صورة عدن على هذا الوضع والترتيب

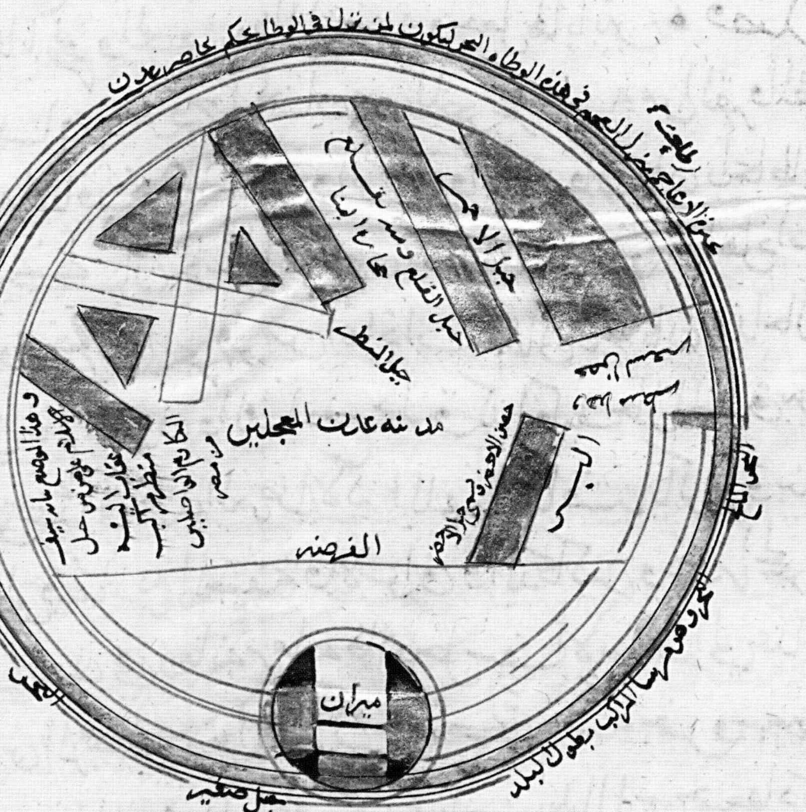

صفة عدن وذكرها بنا البلد في وادي البحر مستدير وحوله هواه كرب ولكنه يقطع حل الحمر في مدة عشرة ايام وماوها من الابار وشي يجلب من مسبرة فرسخين والله اعلم **ذكر الابار العذبة** داخل عدن بير حلقم عود السلطانيه وبير على بن ابي البركات بن الكاتب قديمه وبير احمد بن المسيب وبير ابن ابي الغارات قديمة عند باب عدن وبير المقدم قديمه وثلاثه ابار لداود بن مصنور اليهودي وثلثه ابار للشيخ عمر بن الحسين وبير لعلي بن الحسين الازرق وبير جعفر

house which separated the urban space from the outside world. Within the city walls, however, very little is usually shown: we can only see the palaces of the ruling dynasty in Mahdia and a few labels in Palermo. In reality, these cities were densely cramped within their walls, but the maps do not reflect that. There are no pictorial representations of markets, or of religious and civic institutions. Surprisingly, perhaps, medieval Islamic town views show no mosques.

Why was it so? Why did medieval Muslim map-makers represent the boundaries, walls and palaces of cities but not their internal institutions? The answer may lie in the overt political message of these town views. The urban plans that we have from medieval Islam are associated with political authority, especially over port towns. The city is imagined as a place of refuge, from the sea as well as from the surrounding countryside. This was a result of their intended audience. The author of the *Book of Curiosities* wrote and made maps for a patron in the Fatimid court, maybe for the caliph himself. Such maps did not appeal to religious scholars, who generally avoided visual representations of the urban spaces they inhabited. It is not that men of religion lacked a sense of local patriotism. Since the twelfth century, many Muslim scholars had written volumes dedicated to the history of their home towns and their notable personages. But these local histories and topographies are not illustrated with urban plans. Perhaps some of the religious scholars were piously reticent about the use of images; others may have found city maps to be too closely associated with the political power of military elites.

(previous spread) Map of the port of Aden from Ibn al-Mujāwir's *Tārīkh al-Mustabṣir*, copied 1595. Courtesy of the Turkish Institute of Manuscripts, Süleymaniye Manuscript Library, Istanbul, MS. Ayasofia 3080, fol. 53b.

The Author of the *Book of Curiosities*

The author of the *Book of Curiosities* remains a mystery. He was a loyal subject of the Fatimid state, and the extent of his geographical knowledge suggests a direct connection to the Fatimid state and possibly also to the Ismaili missionary network. His astrological interest in the effects of the celestial sphere on events on Earth, while not completely unusual for the time, also hints at an esoteric Ismaili outlook. The map of Palermo with its suburbs, the diagram of Tinnīs and in particular the map of Mahdia, which is drawn from the perspective of someone looking at the city from a vantage point just outside its walls, suggest that he had visited these port cities in person. But his interest in trade is minimal, and he is more likely to have been a military man than a merchant.

He was primarily a map-maker rather than a scholar. It is the maps that make the *Book of Curiosities* such a distinct work of medieval scholarship and such an appealing manuscript for modern audiences. The author has unprecedented confidence in the ability of maps and diagrams to convey information. Unlike any other geographical treatise beforehand, some of the maps are stand-alone artefacts, unsupported by any accompanying text. This is true of the circular diagram of the skies, the rectangular map of the world and the maps of the three great seas, none of which is attached to an explicatory text. Even when the maps are related to a text, such as those of the islands of Sicily and Cyprus or of the city of Mahdia, the information they contain goes well beyond that of the preceding prose sections. We do not have the original treatise, only a later copy, so we do not know how lavishly illustrated it might have been when it was first penned. But the second part of the title literally translates as 'that which is pleasant to the eyes' (*mulaḥ al-ʿuyūn*), indicating that this treatise is about the images as much as it is about the text.

The anonymous author drew the maps of Sicily/Palermo, Mahdia and Tinnīs to express a political, military message. Ironically, the grandiose depiction of walls and palaces could not prevent the imminent demise of the Fatimid caliphate as a Mediterranean power. These maps may have been useful as pieces of propaganda, but they were no substitute for military might. Drawing thick walls does not stop a Norman onslaught. Within decades of the composition of the *Book of Curiosities*, this trio of Fatimid Mediterranean ports fell in quick succession. Palermo was lost to the Normans in 1072, and the last Muslim stronghold in Sicily capitulated in 1091. The island was never to be recaptured by Muslim armies. The Fatimids lost Mahdia in 1057, when it fell into the hands of a local North African dynasty. It then came under repeated Norman attack, which led to its seizure by Roger II of Sicily in 1148 for a period of twelve years. As for Tinnīs, in 1227 the Ayyubid sultan al-Malik al-Kāmil ordered the destruction of the remaining fortifications so that they would not serve as a base for an invading Crusader army. They are now buried under the sand, with the map of Tinnīs in the *Book of Curiosities* a rare testament to the city's medieval brilliance.

ملندر
جزيره سقوطرى
مرباط
شرمه
حاسك
شبام
حضرموت
بلاد عاد

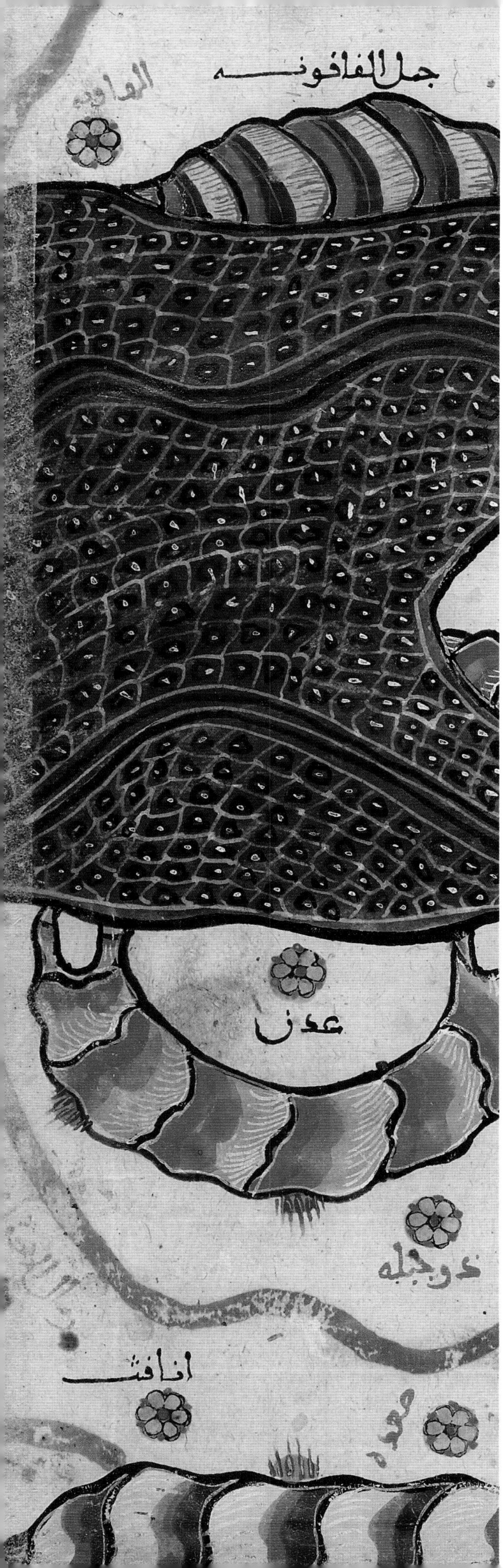

CHAPTER 4

The Grid of al-Sharīf al-Idrīsī

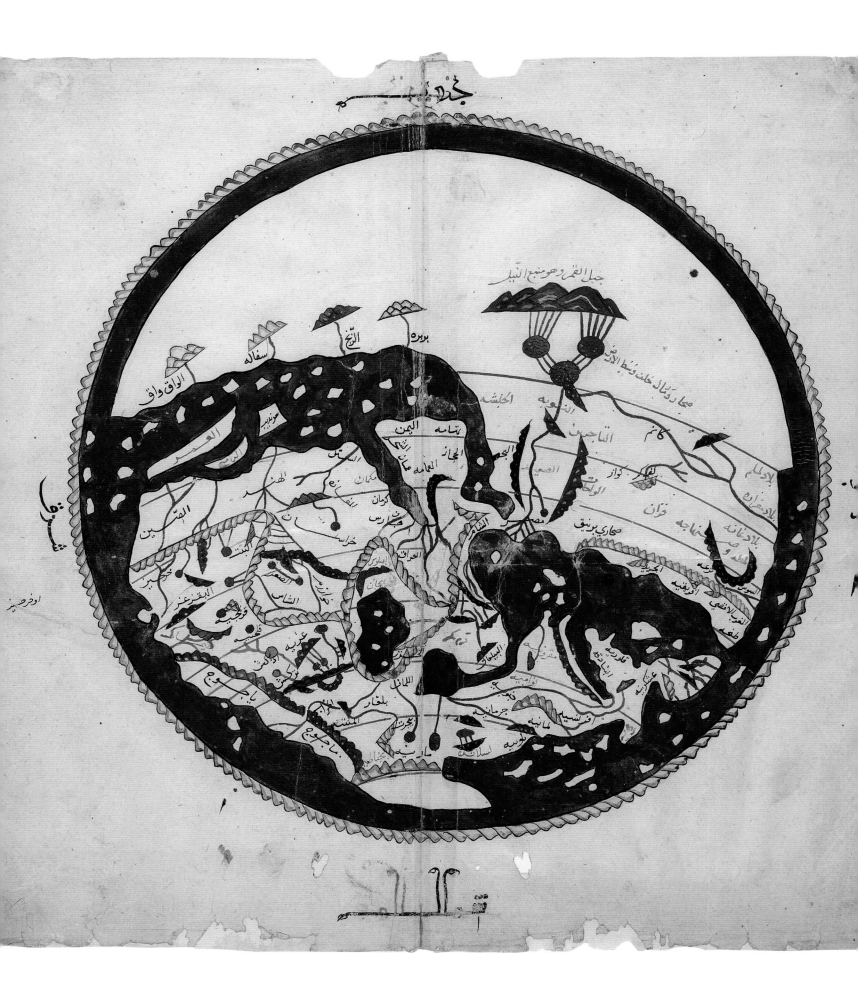

جنوب
جبل القمر وهو منبع النيل
بربره
الزنج
سفاله
الواق واق
تهامه
اليمن
الحجاز
اليمامه
عمان
الحبشه
النوبه
كانم
الهند
السند
فارس
خراسان
الصين
العراق
الشام
مصر
صحاري برنيق
المغرب الاقصى
افريقيه
بلغار
ياجوج
ماجوج
جرمانيه
شمال

The world map of al-Sharīf al-Idrīsī, from *Entertainment for He who Longs to Travel the World*, copied 1553. South is at the top. Bodleian Library, University of Oxford, MS. Pococke 375, fols 3–4.

THE MOST ICONIC AND widely reproduced map from medieval Islam is the world map that accompanies the *Entertainment for He Who Longs to Travel the World*, a geographical treatise by al-Sharīf al-Idrīsī, who worked at the court of Roger II, the Norman king of Sicily. The map shown here is found in a manuscript made in North Africa in the sixteenth century, one of two copies of the *Entertainment* held in the Bodleian Library (opposite). In its layout, this is not substantively different from the world map of al-Iṣṭakhrī. It is a circular representation of one hemisphere surrounded by the Encompassing Sea, with south at the top, and an enlarged African continent that extends eastwards in parallel to the Asian land mass. But, while al-Iṣṭakhrī's world map was an abstract diagram of straight lines with boxed boundaries for provinces and regions, the curves of the coastlines on this map are much softer and more refined. There are also more details, especially in the lands beyond Islam. The shape of Europe, which al-Iṣṭakhrī presented as a perfect triangle, is here much more precise than ever before, with the Italian and Balkan peninsulas well defined. There are also seven red clime lines, starting from the equator and descending towards the north, indicating the Ptolemaic latitudinal divisions of the inhabited world. Otherwise, the world here is an integral whole, with the labels guiding the viewer without imposing political or ethnic divisions on the physical space. Carefully executed mountain ranges and rivers dominate the map, which is reinforced by a palette of colours that culminates in the deep blue of the sea.

Al-Idrīsī is nowadays somewhat of a celebrity. Not only does this circular world map adorn many an introduction to Islamic history, but his own biography has become a symbol of European–Islamic scientific collaboration. Al-Idrīsī is the main character in *A Sultan in Palermo*, a widely acclaimed historical novel by Tariq Ali, a work of fiction with direct topical references to the place of Islam in contemporary European societies.[1] A widely used geographic information system (GIS) for the modelling and analysis of complex geographical data is named IDRISI, in honour of his achievements. His textual description of Europe has been translated several times into French under the title *The First Geography of the West*; here, too, reference to the religious identity of the author carries an implicit topical message.[2] And al-Idrīsī is also the only representative of the Islamic map-making tradition in the recent *History of the World in Twelve Maps* by Jerry Brotton, who views al-Idrīsī as a striking example of east–west exchange, ultimately doomed but nonetheless instructive.[3]

(following spread) The fourth section of the first clime, showing the sources of the Nile, from al-Sharīf al-Idrīsī's *Entertainment for He who Longs to Travel the World*, copied *c.*1400. Bodleian Library, University of Oxford, MS. Greaves 42, fols 17b–18a.

Political undertones aside, the accolades heaped on al-Idrīsī are absolutely justified. Al-Idrīsī was the most ambitious cartographer and geographer of the Middle Ages, synthesizing classical, Islamic and European sources. The highly original way in which he created sectional maps, each corresponding to a uniformly sized square in a grid of the inhabited world, resonates with the digital age. He was also a reflective map-maker, who clearly explained the expected function of his maps and his working methods. His status as a Muslim scientist working, in Arabic, at the court of a Christian monarch is unique among the geographers of the medieval Islamic world, leading him to take a remarkably universalist outlook. Al-Idrīsī should be celebrated.

Most modern biographies of al-Idrīsī tell us that he was born around 1100 in Ceuta, on the North African coast of present-day Morocco, and then educated in Córdoba, the capital of Islamic Spain.[4] We are also often told that he then travelled to Asia Minor, France, England and Spain, before he was invited by Roger II to his court in Palermo around 1138. Roger's motivations in inviting al-Idrīsī are presented as primarily political: he wanted to use al-Idrīsī's claim of descent from the line of the Prophet, as indicated by the title 'al-Sharīf', to bolster his plans for expansion into North Africa. Roger then asked al-Idrīsī to construct a world map and to write a commentary on it; the world map was engraved on silver and was subsequently lost, but the geographical treatise – the *Entertainment* – was completed in January 1154, just a month before Roger's death on 27 February of that year.

Fresh research has corrected key aspects of this standard account.[5] First, al-Idrīsī was almost certainly not born in Ceuta but in Sicily itself, and he grew up within the royal court. According to a little-known fourteenth-century text, only recently come to light, al-Idrīsī's grandfather was the last ruler in a local dynasty that governed Malaga in southern Spain.[6] Following his deposition in 1058, his descendants dispersed all over the western Mediterranean. One of them, al-Idrīsī's father, arrived at the Norman court of Sicily sometime before 1090, and was generously received by Roger I, predecessor to Roger II. According to that fourteenth-century biography, al-Idrīsī himself then grew up in the company of the future monarch. This revisionist biographical narrative fits better with the evidence from al-Idrīsī's own work, where there is no mention of being called from abroad to Roger's court or of any travels in Asia Minor, France or England. If al-Idrīsī ever left Sicily, it was probably only to travel to Muslim North Africa and Spain. We know that he remained attached to his ancestral homeland of al-Andalus – in a separate work, an influential botanical compendium on plants and their medical uses, he kept referring to the plants that grow 'in our land' of Spain (*al-Isbāniya*).

Al-Idrīsī, then, was not some unknown prince randomly plucked from the Muslim Mediterranean. Most likely, he was born into the multifaith, trilingual elite that gathered around the Norman kings of Sicily, who created a unique liminal space between Latin Christendom and Islam. When the Normans seized the island from the Fatimids in 1068–71, they found a mixture of Greek and Arabic; instead of replacing that mixture, they decided to add to it. At the Norman court, al-Idrīsī was surrounded by many Greeks occupying high official positions, including George of Antioch, the most important general and diplomat of the period. The influence of Arabic and Islamic civilization was even more pronounced, as can be seen in the exquisite Capella Palatina in Palermo, a Christian building that draws heavily on the distinctive Fatimid artistic style. Through most of the twelfth century Arabic was the main language of the royal administration, known by its Arabic name Dīwān. Many administrative decrees were issued in all three languages, Greek, Arabic and the rulers' Latin.

Sicily is a Mediterranean hub by merit of its geographical location, and we have seen what an important asset it had been for the Fatimids of Cairo. After its capture by the Normans, and with the advent of the first wave of the Crusades in 1096, it became even more important as a stopping point for Latin fighters on their way to the Holy Land. At the same time, it was also turning into a place of intellectual exchange, second only to the Iberian Peninsula. From the 1140s onwards, Roger actively sponsored several translations of Greek and Arabic texts into Latin. Between 1154 and 1160, works by Aristotle and Plato were translated by Henry Aristippus, former archdeacon of the Church of Saint Agatha in Catania. Henry brought with him from Byzantium a copy of Ptolemy's *Almagest*, which was also translated into Latin, as was the most famous Arabic star catalogue by ʿAbd al-Raḥmān al-Ṣūfī. The compendium of medical plants that al-Idrīsī authored was multilingual, with synonyms recorded across several languages. Although it was written in Arabic, the geographical work can also be seen as a form of translation, as a transfer of Arab–Islamic geographic knowledge and methods to a Latinate context.

In his introduction to the *Entertainment*, al-Idrīsī credits Roger with initiating the project of mapping the world. Roger, 'whom Allah glorified, king of Sicily, Italy, Lombardy and Calabria', and who was knowledgeable in mathematics and mechanics, wished to grasp the geography of the lands that had come under his rule:

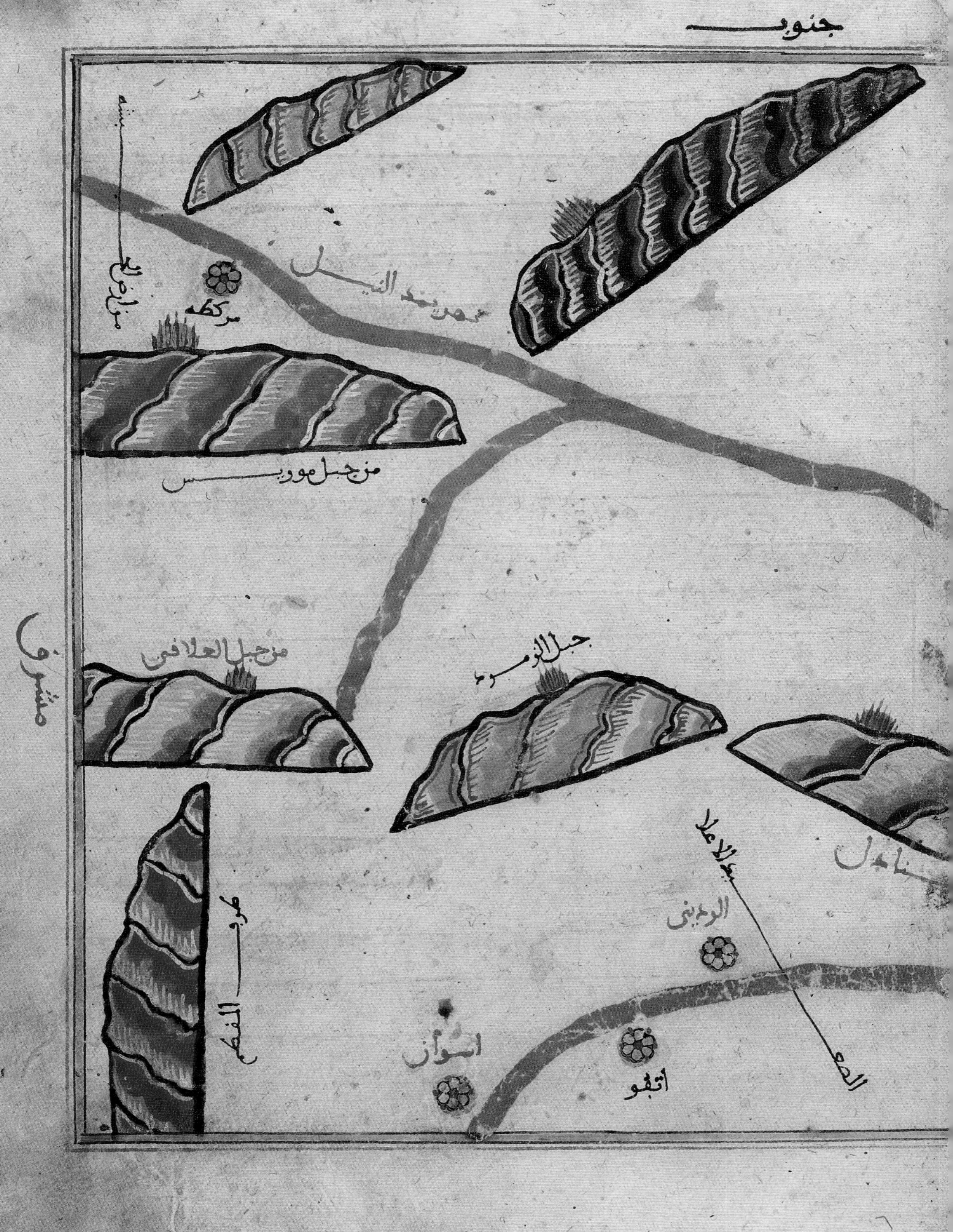
جنوب
مشرق
مركطه
من جبل موريس
من جبل العلاقي
جبل الزمرد
اسوان

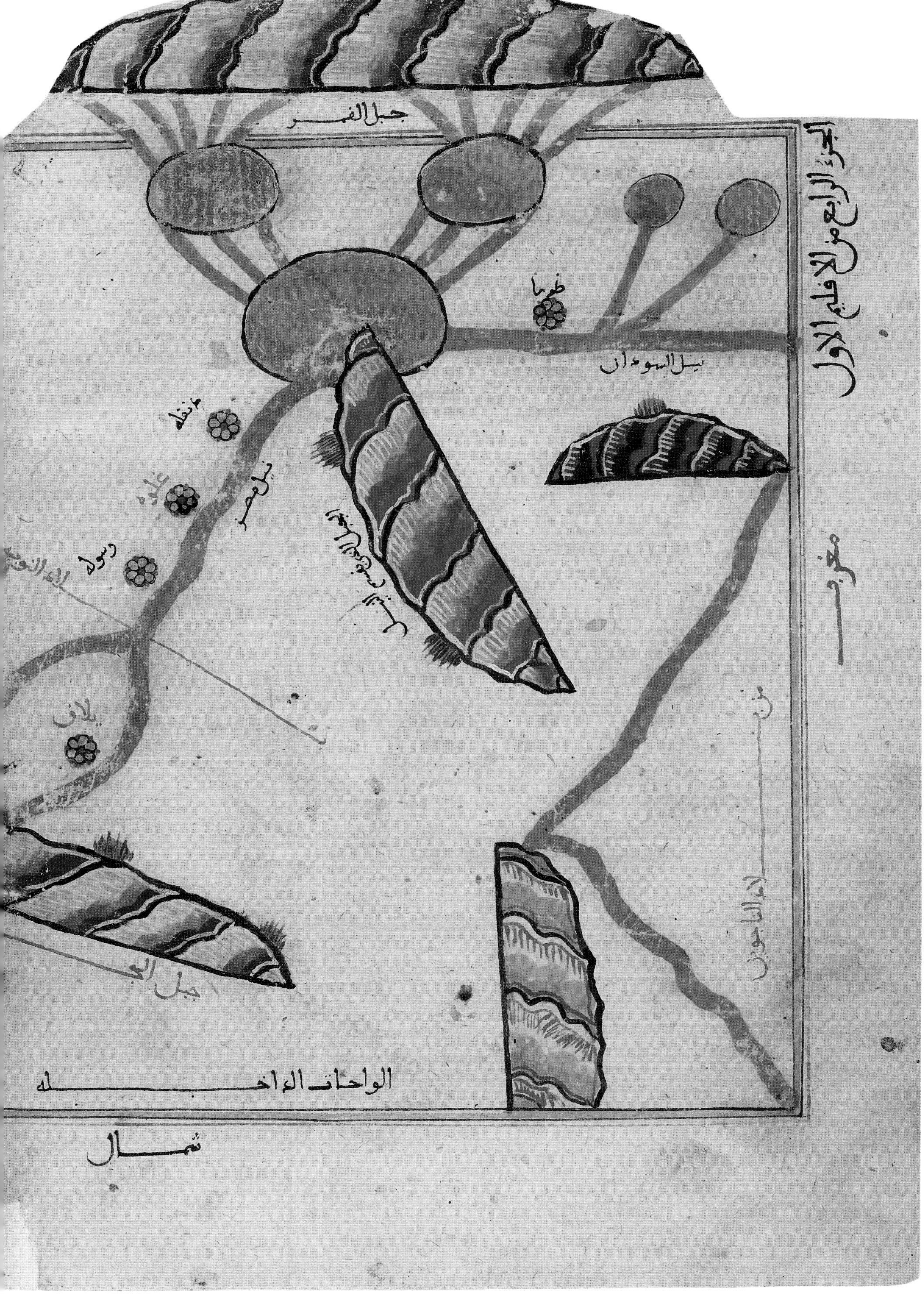

جبل القمر
الجزء الرابع من الاقليم الاول
نيل السودان
دنقلة
علوه
نيل مصر
يلاق
الواحات الداخله
مغرب
شمال

When the provinces of his land expanded and the ambitions of his court increased, and as the lands of the Europeans and their people came under his authority, he wished that he should know accurately the details of his domains and master them with definite knowledge; and that he should know the boundaries and the routes of his domains, both by land and by sea; and in which clime they were located; and the seas and bays that are found in them; as well as knowledge of other lands and regions in the seven climes and the countries that belong to each clime, as much as has been agreed upon by knowledgeable informants and confirmed in the ledgers of transmitters and authors.[7]

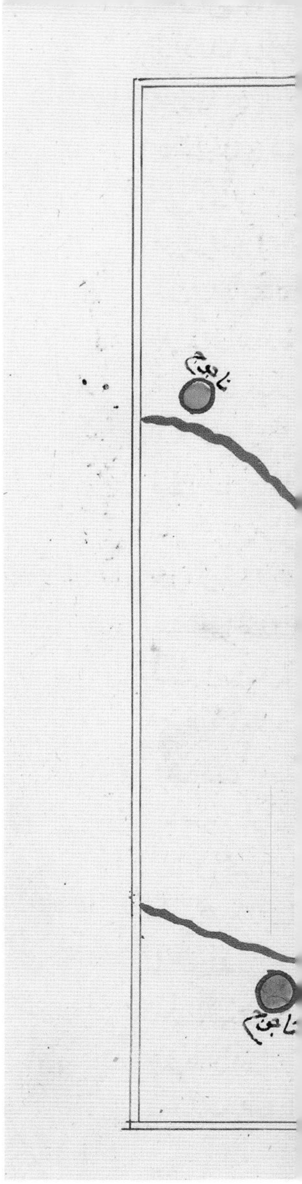

According to al-Idrīsī, the initial impetus for the project was the expansion of Roger's domains. Towards the end of his reign, Roger captured some new territories, first by entrenching his authority in Calabria in the face of papal excommunication, then by taking several towns on the North African coast, including the ancient Fatimid capital of Mahdia in 1148. Despite this expansion, his kingdom was by no means a global empire. Yet, in al-Idrīsī's account, Roger's ambitions appeared to shift unexpectedly from a project of mapping his own lands to one of mapping the world. In order to find accurate knowledge of all seven climes, Roger gathered books and scholars together but was dissatisfied with them, so, in the manner of the rulers of great empires, he sent travellers to all the corners of the world. The fourteenth-century account by al-Ṣafadī adds that each traveller was accompanied by a draughtsman, who drew the forms they observed with their own eyes.[8]

The purpose of collecting the geographical data was to create a map of the world, first as a draft, then engraved in silver. Al-Idrīsī tells us that Roger decided to collate the data on the distances between lands by using a drawing board, over which the localities of the world were placed by using precise measurement tools made of iron. Once Roger was satisfied with the map produced on the drawing board, he had it transferred to another medium:

He ordered that a large and heavy disc should be produced in pure silver, weighing 400 Roman ratls – each ratl of 112 dirhams [1 dirham = 3 grammes]. When it was ready, he ordered the craftsmen to engrave on it the image of the seven climes, with their lands and regions, their shorelines and hinterlands, gulfs and seas, watercourses and rivers, inhabited and uninhabited parts; as well as the frequented routes, the distances in miles, the confirmed lengths of journeys and the well-known harbours between one

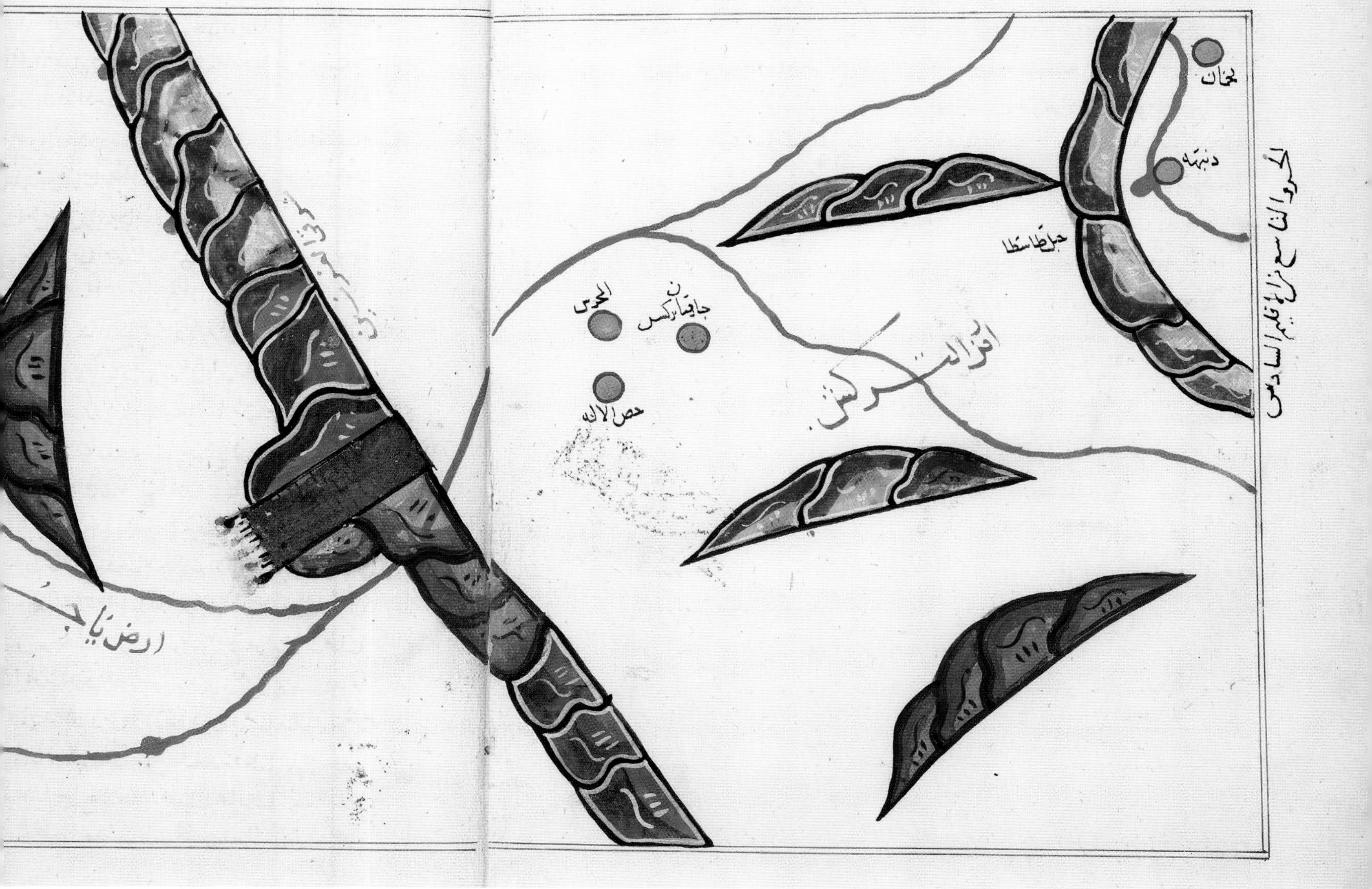

The ninth section of the sixth clime, showing Central Asia and Siberia, from al-Sharīf al-Idrīsī's *Entertainment*, copied 1553. Bodleian Library, University of Oxford, MS. Pococke 375, fols 304–5.

(following spread) The sixth section of the first clime, showing the Gulf of Aden, from al-Sharīf al-Idrīsī's *Entertainment*, copied *c.*1400. Bodleian Library, University of Oxford, MS. Greaves 42, fols 26b–27a.

> locality and another. He ordered that this should be done exactly according to the information on the drawing board, without diverging in the least from its form and structure, as he had drawn it for them.[9]

This map of the world engraved on a heavy silver disc did not survive. Al-Idrīsī does not say anything about its eventual use, but it must have been intended for display purposes. Other rulers, both in Europe and in the Islamic world, had maps made on expensive materials to adorn their palaces. One such world map, made in 964 for the Fatimid caliph al-Muʿizz, was produced on a piece of high-quality cloth woven with gold. The fifteenth-century author al-Maqrīzī stated that 'it had the image of the climes of the Earth, with its mountains, seas, cities, rivers and routes, similar to the *Geographia* [of Ptolemy]. It also had Mecca and Medina prominently

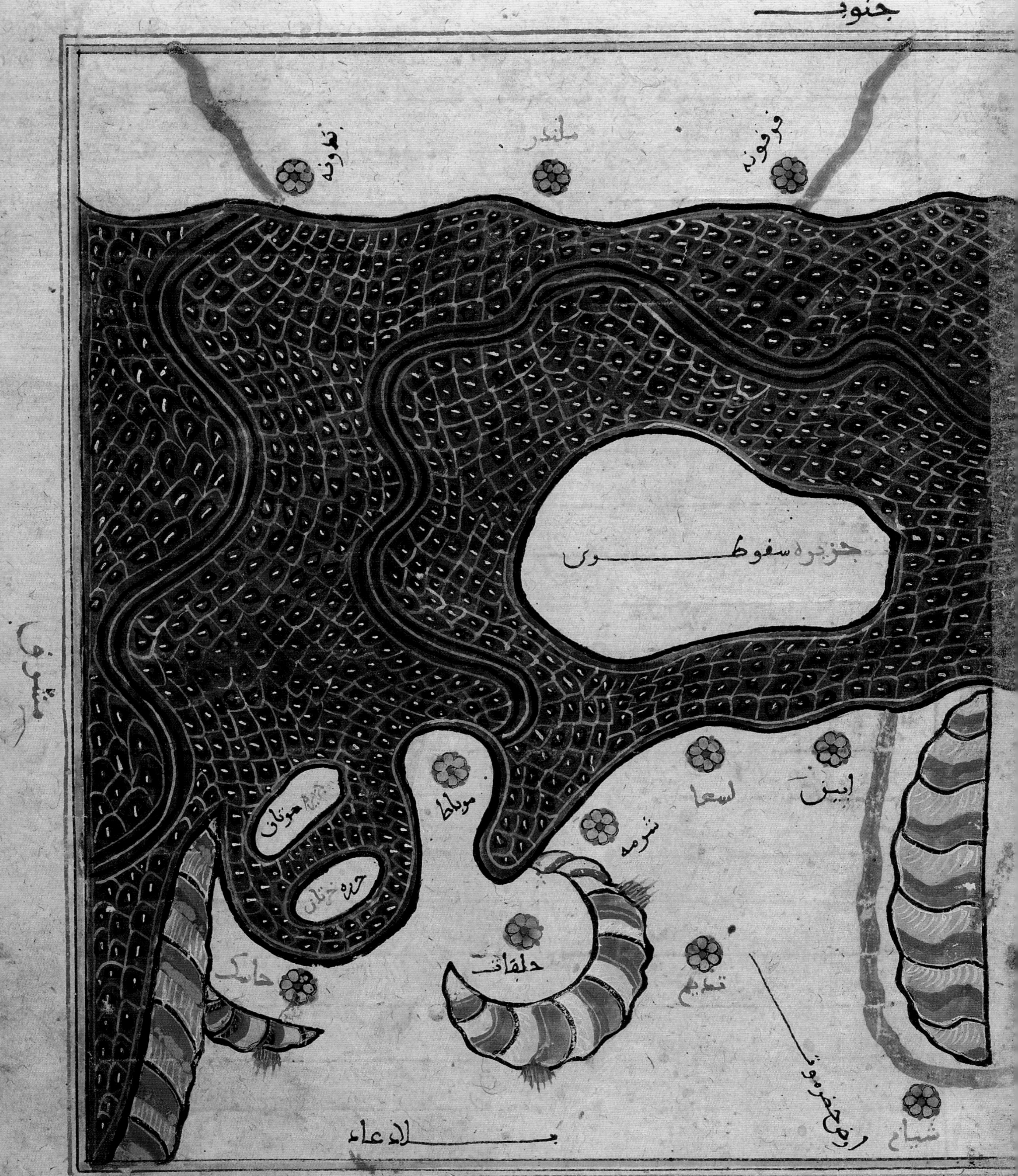
جنوب
مشرق
ملندر
جزيره سقوطرى
مرباط
شرمه
لسعا
ابين
حاسك
شبام
رمل حضرموت
بلاد عاد

الجزء السادس من الاقليم الاول

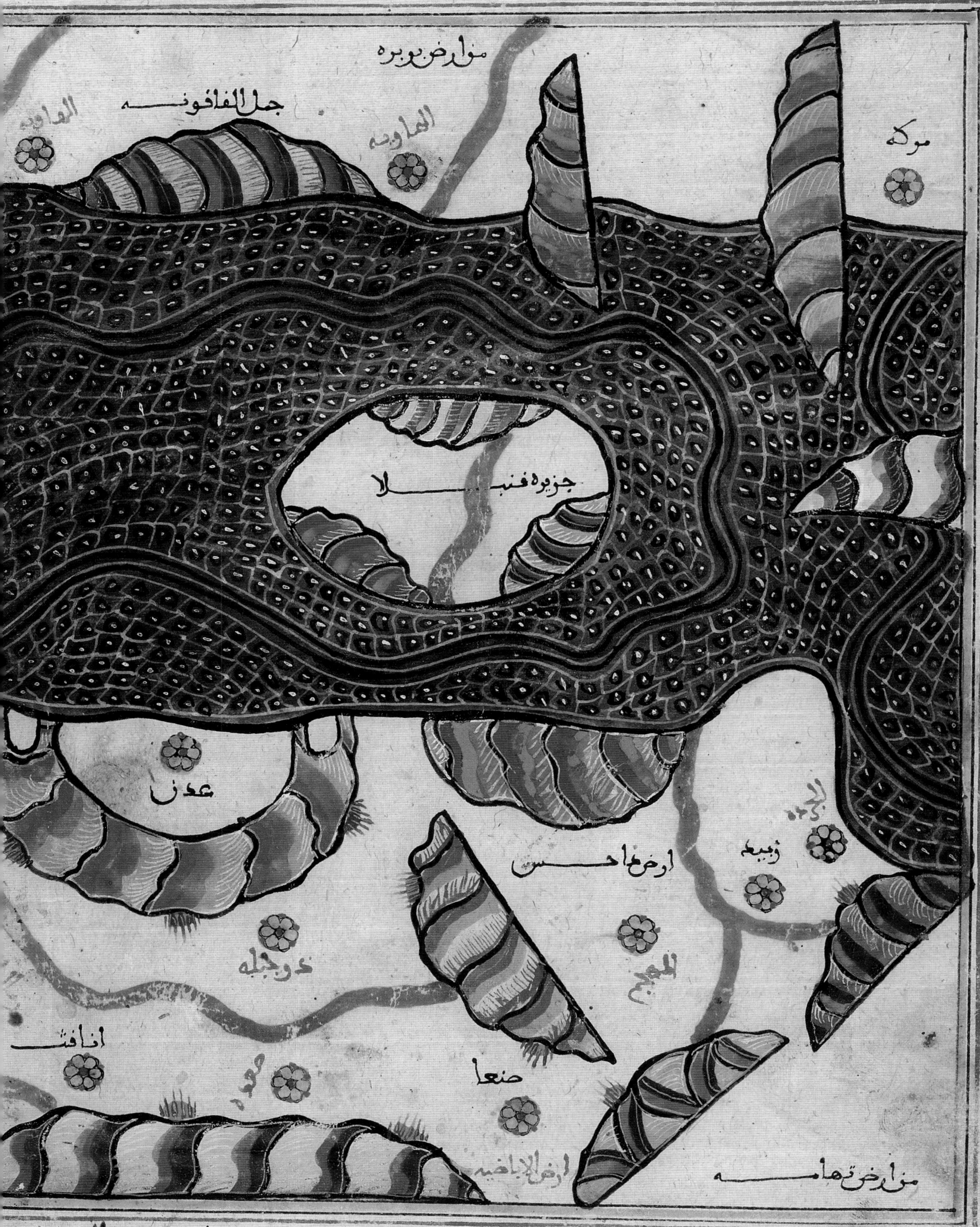

مغرب

شمال

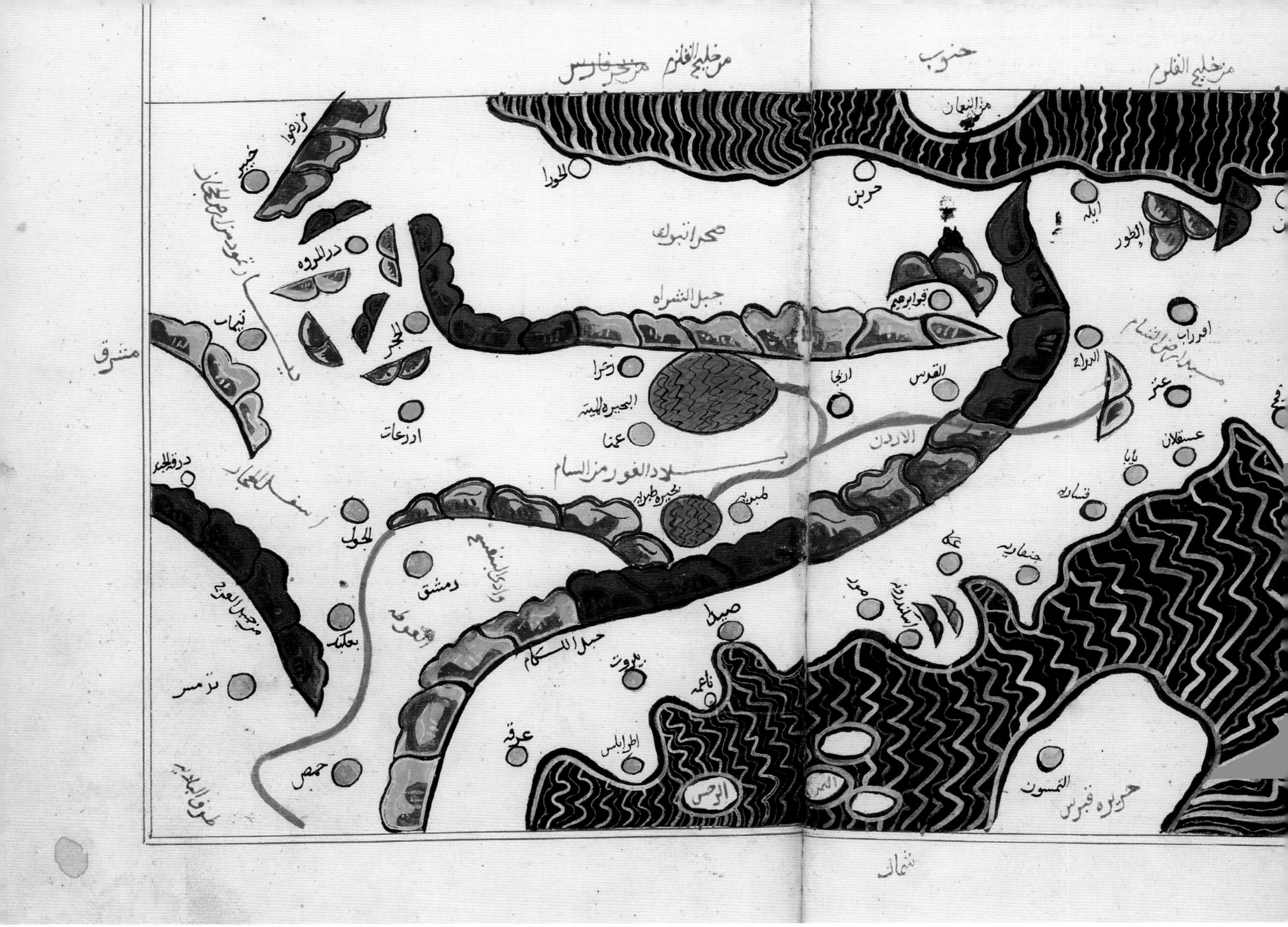

indicated. Each city, mountain, land, river, sea and route had its name written in gold, silver or silk.'[10] This Fatimid world map did not survive either; it was taken by rebelling troops who looted the caliph's palace in Cairo in 1068.

Al-Idrīsī's iconic circular world map is, most likely, a miniature replica of that lost silver engraving made for Roger. The introduction to the *Entertainment* doesn't mention or explain this circular world map, and it is found in only some of the surviving copies of the treatise. Revisionist scholarship has even suggested that the circular world map that has reached us may not be al-Idrīsī's at all, and that it came from yet another map-making tradition unknown to us. This is supported by the surprising presence of a copy of the same circular world map in the recently discovered Fatimid *Book of Curiosities*, written in the eleventh century

The fifth section of the third clime, showing Palestine and Syria, from al-Sharīf al-Idrīsī's *Entertainment*, copied 1553. The map is oriented to the south, but Palestine is shown as an elongated band on an east–west axis. Bodleian Library, University of Oxford, MS. Pocock 375, fols 123b–124a.

and copied around 1200. But the balance of probability is that the circular world map is an integral part of al-Idrīsī's work. Like the silver disc, it highlights the seven climes, drawn here with strong red lines, as well as the physical topography of coastlines and rivers. There are also distinctive aspects of the circular world map that link it to the rest of the treatise, such as the increased precision of the European coastlines and a western arm of the Nile that flows towards the Atlantic Ocean. Moreover, if we assume that the world map did originate with the silver disc, we can also account for the refinement of its lines and the harmony of the image; this was a map made for a king.

After the engraving of the silver disc was completed, Roger commissioned al-Idrīsī to write a book that would explain the world map and expand on it. Al-Idrīsī tells us that this happened in Shawwāl 548 of the Islamic calendar, or January 1154, that is, a month before Roger died. It seems that the commissioning of the *Entertainment* was one of the final requests of a dying man. According to al-Idrīsī's own account, it was only then that his direct involvement in the project started. Roger had initiated the project of mapping the world, traced it on a drawing board and called for it to be engraved, probably for display. On his deathbed, he asked al-Idrīsī to write a book that would capture all that a single world map could not.[11]

With a world map engraved on silver to hand, al-Idrīsī was faced with the problem of zooming in on the details of every inhabited region – the problem facing all atlas-makers. Most, like al-Iṣṭakhrī before him, had chosen to do so by dividing the world into physical units, such as continents or political units. Al-Idrīsī's highly original solution was to disregard physical or political boundaries, and instead to divide the world into uniformly sized squares derived from a notional grid. His starting point was the the division of the inhabited world into seven climes, or latitudinal bands, as found on the world map. Since each clime stretched from east to west, covering the entire 180 longitudinal degrees of what was considered to be the inhabited world, they were still too large for coherent presentation. So al-Idrīsī further divided each clime into ten longitudinal sections, each representing exactly 18 degrees, starting from the westernmost section of each clime and moving eastwards. The result is that each of the seven climes is represented by ten sectional maps, adding up to a total of seventy maps, each of uniform length and breadth, and each followed by a textual account to complement and expand on the visual representation.

This uniform, modular method of representing the world is grounded in mathematical geography, and al-Idrīsī is explicit about his debt to Ptolemy. The

Geographia is listed as his primary source for understanding the physical features of the Earth and its place in the universe. In fact, al-Idrīsī did not use Ptolemy's original work but an adaptation, very likely the one used by al-Khwārazmī in Baghdad several centuries earlier. But, while his framework is derived from mathematical geography, al-Idrīsī largely dispensed with indicating latitude and longitude coordinates, and certainly did not plot his maps.[12] Instead, al-Idrīsī was interested in reconciling the Ptolemaic framework, derived from a work that was now 1,000 years old, with more recent information gathered by the extensive corpus of the Arabic–Islamic geographical writings of the preceding centuries. He seems to have read nearly all of them, not only those who were very familiar with the Mediterranean like Ibn Ḥawqal, but also authors from the far eastern edges of the Islamic world. And, as Brotton rightly says, al-Idrīsī borrowed from everyone but always reached his own conclusions.[13]

A remarkable example of al-Idrīsī's synthesis of the Ptolemaic framework with Islamic-era information is his representation of the sources of the Nile, shown in his map of the fourth section of the first clime (pp. 98–9). As in all seventy sectional maps, the boundaries of the map are the boundaries of the grid. Since this is the first clime, the equator is at the top of the map, which is oriented south, while the line marking the northern boundary of the first clime is at the bottom. The eastern and western boundaries of the longitudinal section are to the left and to the right, respectively. The most prominent and familiar element on this map is the Mountain of the Moon, which lies south of the equator and therefore outside the grid. Also present here is the tripartite lake system that dominates al-Khwārazmī's maps of the Nile, as well as an eastern tributary (coming from the left), whose confluence with the main branch of the Nile creates a large island. All of these features are derived from late antique, pre-Islamic material.

There are, however, two novel and interrelated elements in al-Idrīsī's representation of the sources of the Nile, both located at the top right of the map. One is that the lake at the origin of the Nile is the source of another river, an arm of the Nile that flows not towards Egypt but towards West Africa. The other novel feature is a large diagonal mountain protruding into the lake at the source of the Nile, labelled the 'Mountain that splits the Nile'. Neither of these elements was attested as such in earlier sources, and they appear to be al-Idrīsī's solution to conflicting accounts of the sources of the Nile. On the one hand, al-Idrīsī remained committed to the Ptolemaic model of a Mountain of the Moon at the source of the Nile. But he was also aware of Arabic geographers who reported that a western tributary of the Nile, flowing from sand dunes in West Africa (in reality, this was a result of an ancient misconception of the course of the Niger River). He was also aware of indigenous Egyptian traditions that the mountain at the source of the Nile was not the Mountain of the Moon at all but another mountain whose melting snow fed the Nile every spring.

Al-Idrīsī did not discard any of these reports but adapted them to find a systematic solution to the contradictions they posed. The snow-covered mountain reported by Egyptian informants is transformed from a source of the Nile into a barrier that splits the Nile at its origins. The mountain bifurcates the Nile into the western Nile, Nīl al-Sūdān, in the west and the main Nile to the north. The western tributary of the Nile, which Arab geographers reported as emerging from sand dunes in West Africa, is transformed into a western arm of the Nile, one that flows from central Africa towards the west. This new conception of the sources of the Nile would transform medieval understanding of the entire African river systems. If we look back at the circular world map, we can see clearly how the diagonal mountain range splits the lake at the origins of the Nile into Egyptian and West African arms. This imagined western arm of the Nile was al-Idrīsī's distinct cartographic invention, and would influence Islamic and European maps for centuries.

Another example of the mixing of new and old geographic material comes again from the edges of the inhabited world, this time the north-eastern regions of Central Asia and Siberia, represented in the ninth section of the sixth clime (pp. 100–1). The section is dominated by the barrier constructed by Alexander the Great to enclose the people of Gog and Magog. These mythical nations are localized by the two labels at the far left, or east, of the map, near the banks of two rivers that flow towards the barrier. Al-Idrīsī is here again indebted to pre-Islamic traditions, specifically to the Alexander romance. But on the other side of the wall, on its western side, the label refers to the very real Turkish Ghuzz tribes who inhabited the eastern steppe. While much of the eastern steppe and Siberia are blank, as one would expect, al-Idrīsī is able to fill some of the space with ethnic groups and exotic localities. As with his representation of the origins of the Nile, this sectional map is a zoomed-in version of the relevant parts of the circular world map. Looking back at the world map, we can see the people of Gog and Magog enclosed in the north-eastern corner of Asia (bottom left), separated from humanity by an imposing mountain range.

(following spread) The second section of the sixth clime, showing northern France and the southern coasts of England, from al-Sharīf al-Idrīsī's *Entertainment*, copied 1553. Bodleian Library, University of Oxford, MS. Pococke 375, fols 281b–282a.

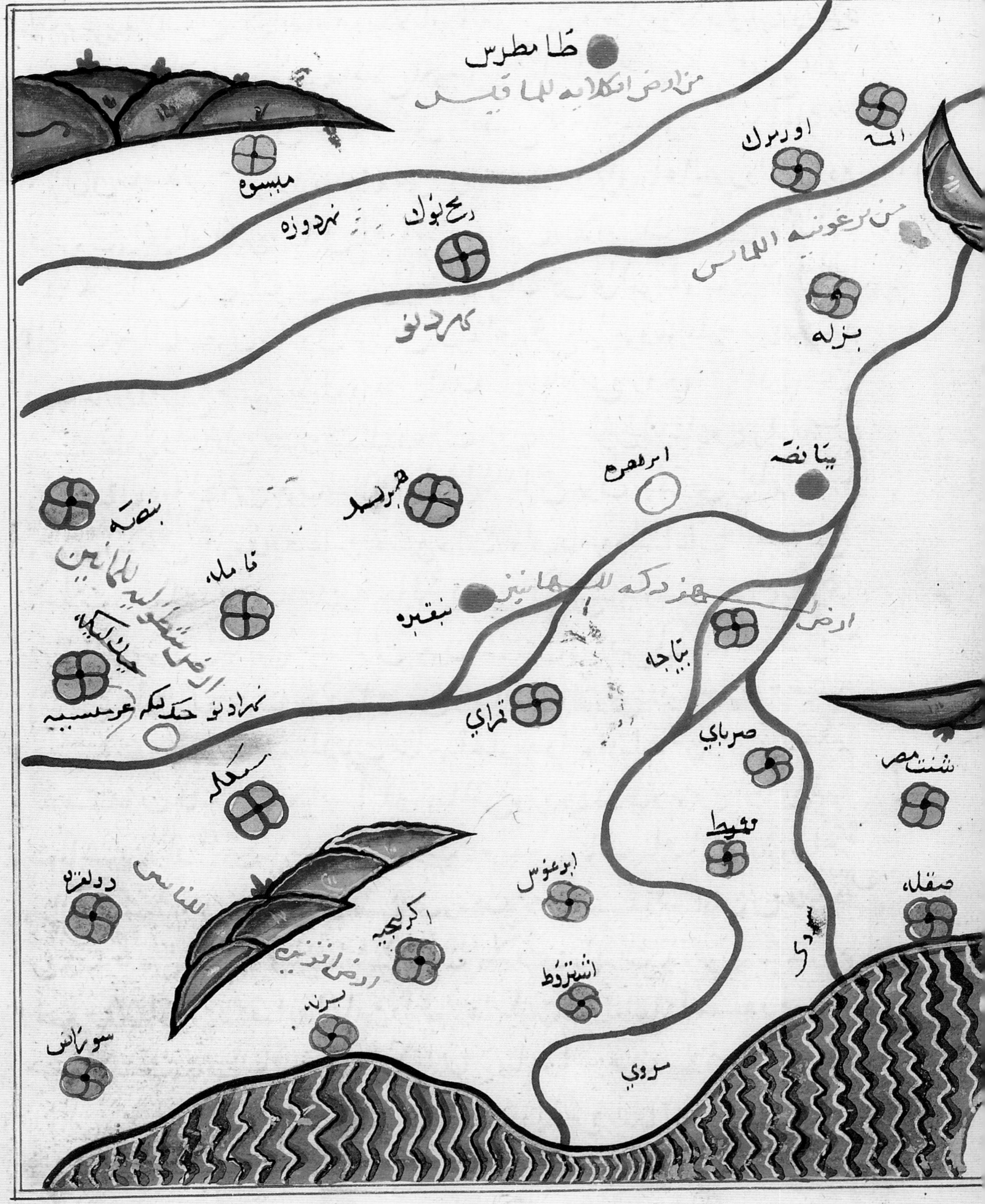
طا مطرس
نهر دوزه
نهر دنو
بيانصه
قامله
صربابي
شنت مص
صقلة
ابو عوس
اشترط
سودي

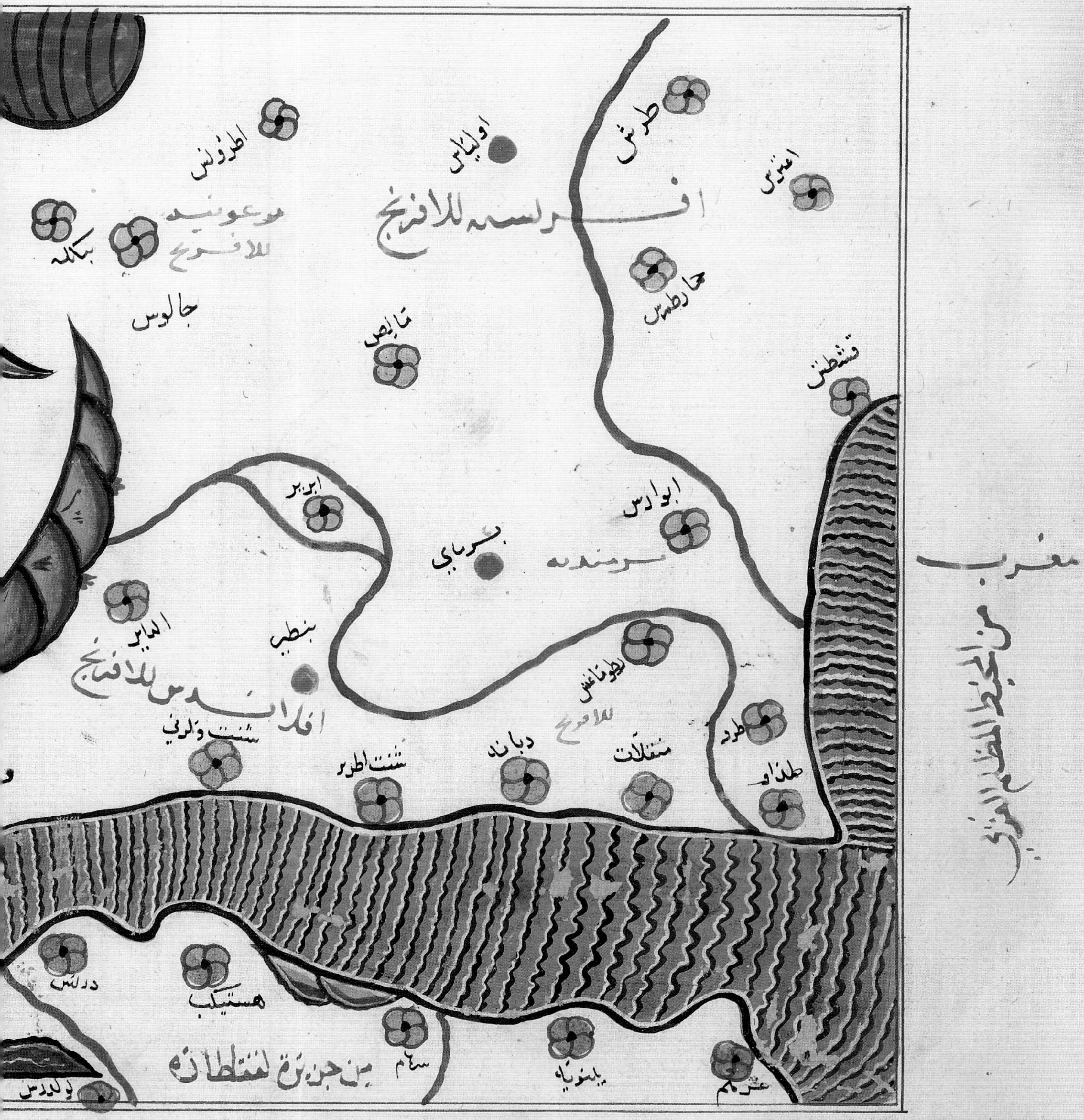
افرانسه للافرنج
جالوس
ابوارس
دباناد
منقلات
مغرب

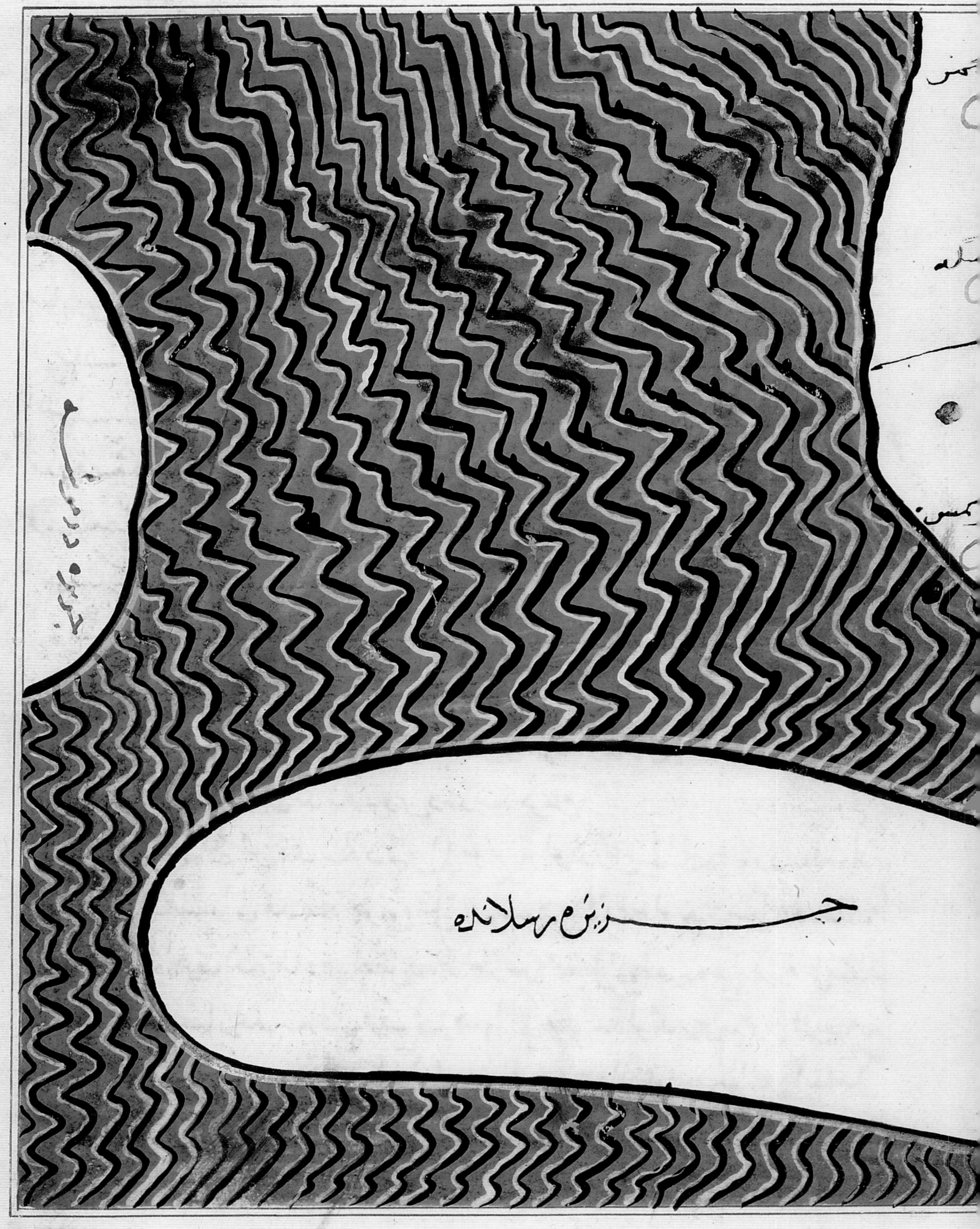

الجزو الثالث من الاقليم السابع

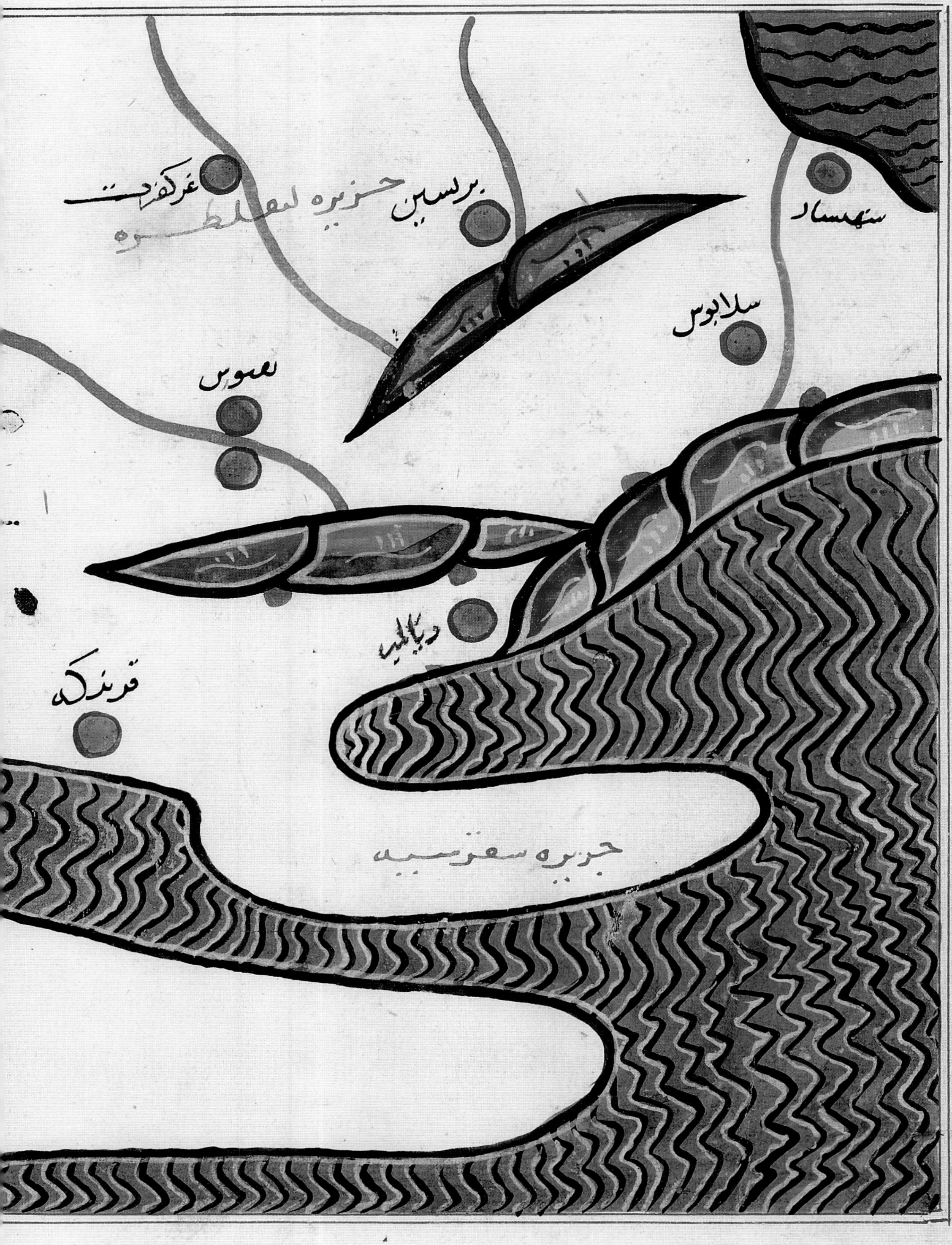

Al-Idrīsī was on firmer ground as he moved into more temperate zones, and was particularly attuned to the representation of maritime space, as shown in the beautiful map of the Gulf of Aden (pp. 102–3). This is the sixth section of the first clime, with the equator at the top of the map. It shows the East African coast in the upper part, facing the southern coast of Arabia in the lower half. The most striking feature on the map is the red semicircular mountain range that encircles the port of Aden on the Arabian coast, slightly to the right of the centre of the map. We have seen how the traveller Ibn al-Mujāwir, about a century later, tried to capture the topography of this major commercial entrepôt. Here al-Idrīsī manages not only to convey the protection offered to Aden by its surrounding mountains but also to place it in an Indian Ocean context. The coastlines, especially on the Arabian coast, are winding, indicating peninsulas and little islets. The two major islands shown, Socotra and Pemba, are wrongly placed and oversized, but an attempt has been made to individuate their shape – these islands are no longer simple circles. The sea itself is shown with a pattern of lines, dots and circles, as is typical of most of the extant copies of al-Idrīsī's maps.

A citizen of the western Mediterranean, al-Idrīsī was well versed in the sea. He did not order his maps according to bodies of water, unlike the author of the *Book of Curiosities*, but his maps and the texts that accompany them show an awareness of the importance of sea travel. It is typical that islands in the Mediterranean and the Indian Ocean are both oversized and individuated, and that the coastlines allow for major peninsulas. The text often provides maritime distances, usually in terms of days of sailing. Distances along the coast are given both as straight lines between the two tips of a bay and as the distance measured along the coastline of a bay. Al-Idrīsī also provides information on prevailing winds, and the Arabic names of the winds sometimes refer to the wind roses used by European mariners, such as *shalūq* (*scilocco*) and *al-libāj* (*libeccio*), or to close synonyms in the Italian used by Mediterranean sailors.[14]

Al-Idrīsī was writing the *Entertainment* during the age of the Crusades, when Italian ships brought Latin fighters to capture Jerusalem from Muslim rule. This Muslim author was writing in Arabic at the court of a Christian king who was an ardent Crusader and whose cousins had made a name for themselves in Syria and Palestine. But al-Idrīsī's map of the Holy Land, shown at the centre of the fifth section of the third clime, reveals almost nothing of this religious tension (p. 104). While the map is oriented to the south as usual, the layout here is confusing. Palestine is shown as an elongated band on an east–west axis, sandwiched between the Mediterranean

(previous spread) The second section of the seventh clime, showing northern and western England, from al-Sharīf al-Idrīsī's *Entertainment*, copied 1553. Oxford is shown at the top (south) of the map. Bodleian Library, University of Oxford, MS. Pococke 375, fols 310b–311a.

at the bottom and the Red Sea at the top. The cities along the Mediterranean coast are easily recognizable, with a deep bay indicating the major Crusader port of Acre, facing the island of Cyprus at the bottom of the map. Inland, the smaller lake is the Sea of Galilee, the larger lake the Dead Sea – shown here in green, mistakenly, to indicate fresh water, but in blue to indicate salt water in other copies.

Jerusalem lies next to the centre of the map, to the right of the two lakes, enclosed by mountains, and by the Jordan River. There is nothing in the map to suggest that it is of special religious significance. The only possible reference to the biblical past is the Tomb of Abraham, that is Hebron, indicated just above Jerusalem and closely surrounded by an orange mountain range. In the text, al-Idrīsī neutrally reports the foundation of the city by Solomon, its conquest by the Muslims and the establishment of Masjid al-Aqsa, and its capture in 1104 by the European Christians who remained in possession of it. In the other sectional maps, Mecca is subject to the same uniform iconography. Al-Idrīsī's Muslim identity is never in doubt, but here and throughout he is 'reluctant to endorse Islamic or Christian claims to universal sovereignty'.[15]

Ironically, the only religious centre to be discussed at some length is Rome. Al-Idrīsī's account of Rome accords much respect to papal authority over all of Christendom, which is surprising given Roger's persistent defiance of papal excommunications, let alone the pope's attitude towards Islam. Al-Idrīsī was not the first Arab geographer to write about western Europe but he was by far the most knowledgeable.[16] Most of his information came from travellers and clerics who frequented the Norman court rather than from books. The only Latin source he mentions is Paulus Orosius, a fifth-century Iberian scholar who wrote a geographical survey of the rise of Christianity, a work permeated by biblical narrative that was hardly up to date by the time al-Idrīsī was writing.

Further north, the map of the second section of the sixth clime, representing northern France and the southern coast of England (pp. 108–9), shows fairly accurate information on the layout of the coasts, and the positioning of individual cities in relation to each other is impressively correct. Paris (Ibrīz) is shown in the centre of the right-hand folio as an enlarged island formed by the Seine. On the left-hand folio, Liège (Biaje) is also correctly shown at the confluence of two rivers. The present-day Belgian towns of Tournai, Bruges and Ghent are shown in correct relation to each other. The red script refers to regional ethnic groups. The large red label in the top right refers to the French (Ifransa) from among the Franks, and the red label below, or north, of Paris refers

(following spread) The third section of the fourth clime, showing the island of Sicily, from al-Sharīf al-Idrīsī's *Entertainment*, copied 1553. Bodleian Library, University of Oxford, MS. Pococke 375, fols 187b–188a.

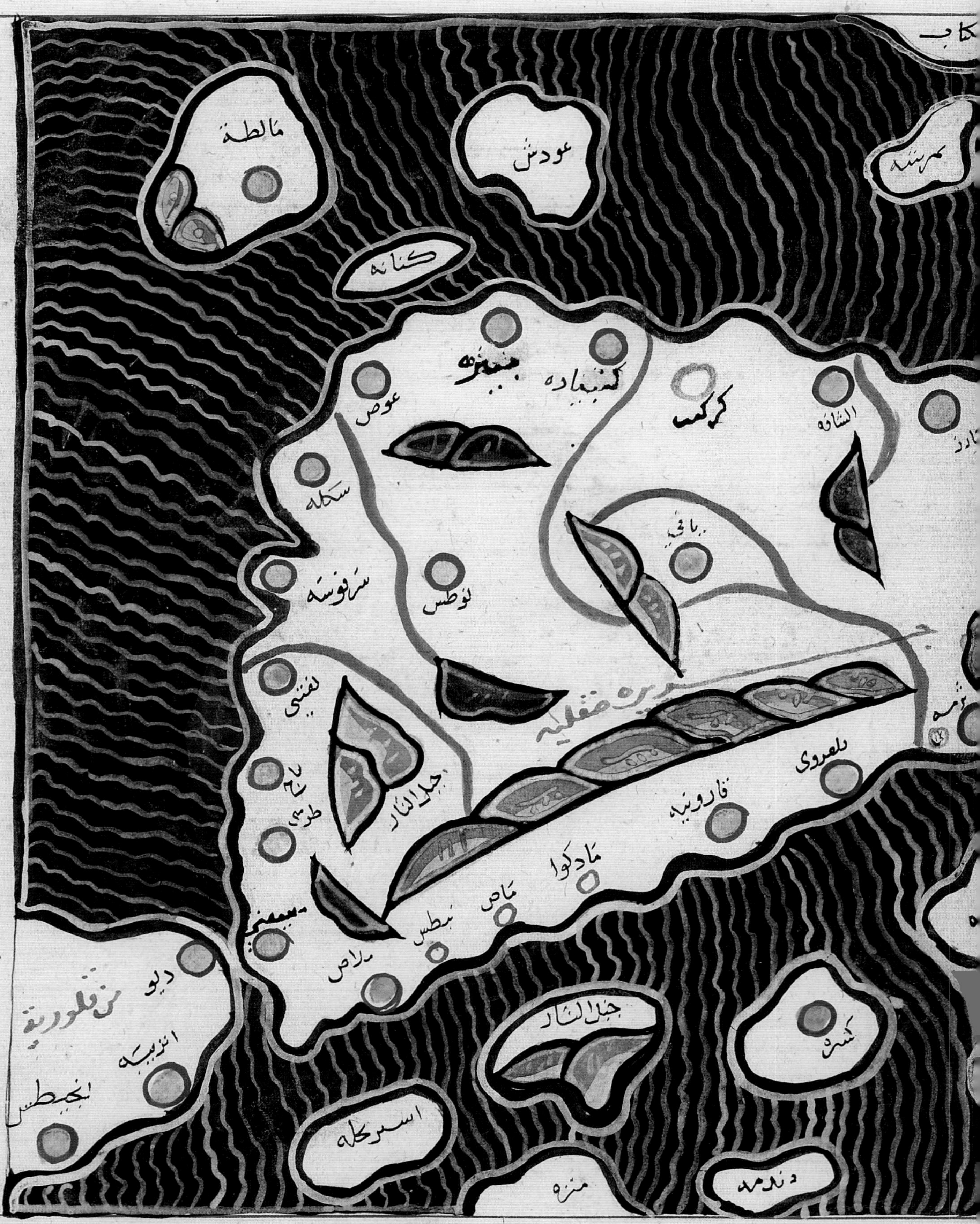
مالطة
عودش
جزيرة صقلية
سرقوسه
لوطس
جبل النار
فاروبه
مادكوا
جبل النار

الجزو الثالث من الاقليم الرابع
فليطمه
اليابسه
القيطيه
سردانيه
برطيز
برشلونه
ربونه
من ارض عشكونيه
جريره
قبره

to the another group of Franks, the Flandres. The red script near Liège refers to the Lotharings from among the Germans.[17]

The southern coast of England is shown at the bottom of the map. Although England was conquered by another group of Normans in 1066, al-Idrīsī's source for his account of the island appears to have been a French-speaking sailor who was familiar with the English coast from Dartmouth to Grimsby but less so with the interior, and who had never visited London or the south-western coast. England, called Inqlitirra (from the French 'L'Angleterre') is described in the text as a fertile island, whose shape 'resembles the head of an ostrich', and whose inhabitants are 'hardy, resolute and prudent'.[18] On the map, the town nearest to the mainland, close to the centre of the bifolio, is of course Dover (written 'Dubris'). A westward sequence along the coast shows Hastings, Shoreham, Southampton (Hantuna) and Wareham. London (Londres) is shown on the bottom edge of the map and described in the text as being 40 miles inland from Dover.

The rest of the island is shown on a map of the second section of the seventh clime (pp. 110–1), found much later in the treatise. The reader who is interested in grasping the correct image of England has to flip many pages in order to get here. This is one of the problems of the uniform grid used by al-Idrīsī, which disregards natural regions and political units. Another problem is the non-economical use of the space on the map. The uniform rectangular grids distort the curvature of the Earth and unnaturally stretch the sections of the already sparsely populated northern climes. Thus, the north is made to seem even emptier than it is, and there is a lot of empty space on this map of the north of England. The strangely formed small peninsula at the lower tip of the island is Scotland (Scosia), an empty space devoid of any settled human habitation. This is also how Scotland is described in the text. The large uninhabited island at the bottom of the map, to the north-east of Scotland, is called 'Rislanda' – perhaps Iceland, or possibly, based on its location on the map, very oversized Orkney Islands. The towns in the west of England, on the right near the green body of water, are Winchester and Salisbury. The red label in the centre of the main island reads 'Island of Inqlitirra'. And the inland town just above the red script, lying on a river that flows south, is labelled 'Ghurkfurt' – undoubtedly Oxford, the small town at the extremity of the inhabited world where the manuscript that contains this map eventually ended up, in the treasures of the Bodleian Library.

Al-Idrīsī was much more familiar with Sicily itself, shown in the third section of the fourth clime with its characteristic triangular shape in the midst of its archipelago, with the toe of Calabria to its left and Sardinia to the right (pp. 114–5). Unlike the map of Sicily in the *Book of Curiosities*, this is not an abstract image but aims to capture the topography of the island as a whole as well as its distinctive shape. The principal rivers and dominant relief are indicated in stylized but recognizable form. Mount Etna, indicated by a purple mountain range, is correctly shown close to the north-east corner of the Sicilian triangle. A similarly shaped purple mountain indicates the island of Vulcano, just below, or north of, Sicily. As is common in al-Idrīsī's maps, islands such as Malta in the top left of the map are oversized and individuated, emphasizing the importance of maritime travel. In Sicily itself, approximately twenty-five towns and cities – all on the coast, except for two in the interior – are marked, invariably as rosettes. Palermo is indicated, but it is one city among many and no attention is given to its fortifications or to the fortifications of any other city. Even Sicily itself, while more familiar to the reader, is nonetheless one island among many, with minimal disruption to the uniform grid through which the world is observed.

It is possible for us to place al-Idrīsī's seventy sectional maps next to each other and to construct a rectangular world map based on his grid. In the modern period, that has been attempted by the German historian of cartography Konrad Miller, who in 1928 created a composite map showing what the sectional maps would look like joined together, based on two manuscripts, one in Paris and the other in the Bodleian Library (overleaf). This composite map, enabled by modern technologies of reproduction, allows us to view the connections between the sectional maps much more easily, and to fully appreciate the richness of the information provided by al-Idrīsī and the sheer ambition of his project. More recently, the Bodleian Library commissioned Factum Arte to create a composite digital scan from the seventy regional maps. But that is not how al-Idrīsī meant us to view his maps. He started his project after Roger's world map was already drawn and engraved on a silver disc, and his mission was not to construct a new one. We should think of the sectional maps as an attempt to expand on an existing image of the world, as if he were methodically going over the circular world map with a magnifying glass, section by section and square by square.

Al-Idrīsī completed the *Entertainment* and its maps sometime around 1158. Roger was long dead and had been succeeded by his son William I, also known as William the Bad, who continued his father's interest in science and promoted several translations of scientific treatises from Greek and Arabic into Latin. There is no indication of how the *Entertainment* was received in William's court. It was

(following spread) A composite map showing what al-Idrisi's seventy sectional maps would look like joined together, based on two manuscripts, one in Paris and the other in the Bodleian Library. Prepared by Konrad Miller, 1928. Bodleian Library, University of Oxford, B1 (135), 3 separate large sheets.

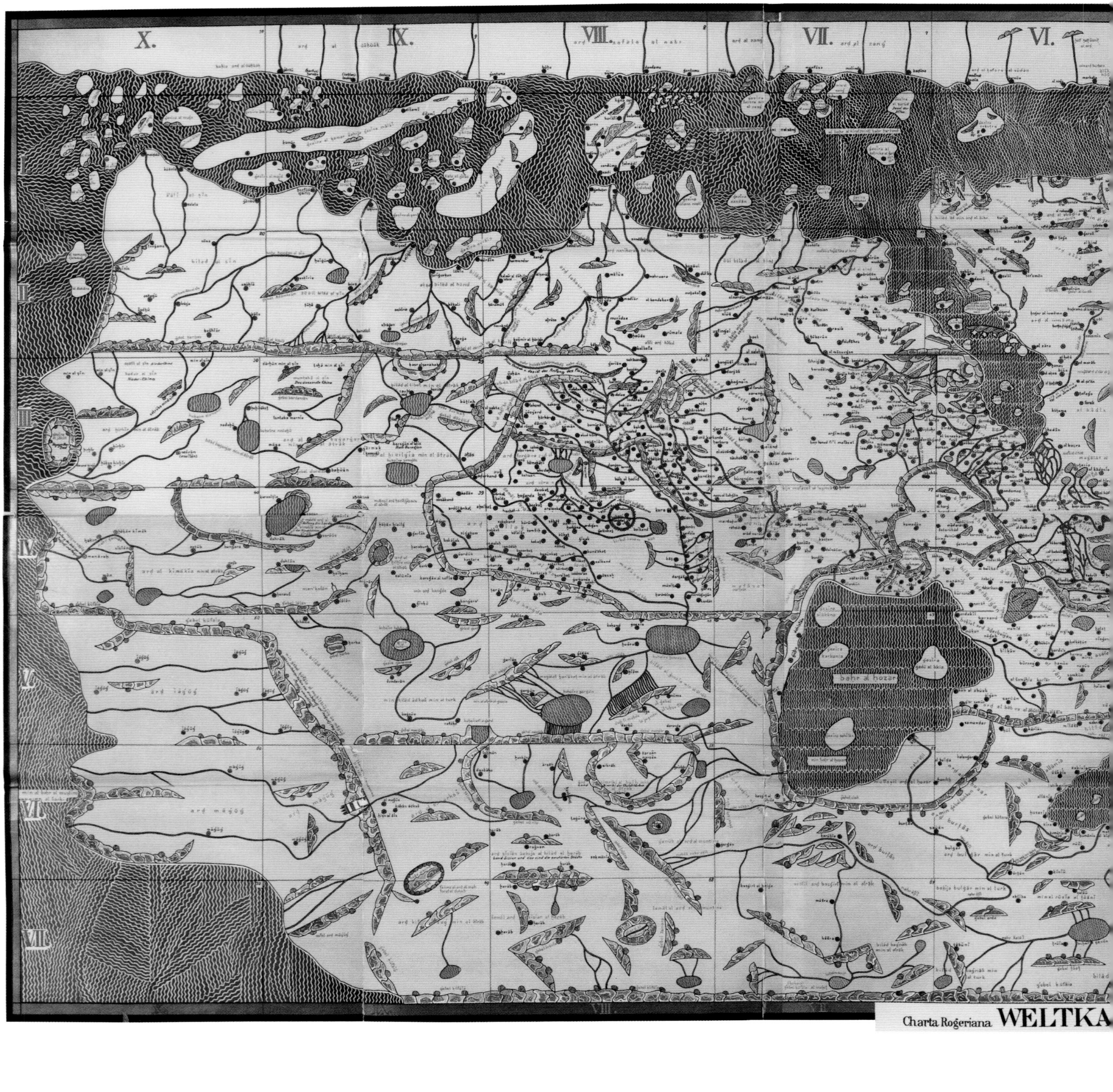

Charta Rogeriana WELTKA

IDRISI vom Jahr 1154 n.Ch.

asia. porrecta ab oriente usque ad meridiem usque ad nilum fluvium extendit. a sept. fluvius tanay et bechium sancti georgii. India a meridiano mari usque ad ortum solis porrigit.
ab occidente Indo fluvio clauditur. a sept. monte caucaso et in ea est mons caspius a quo caspium vocatur mare. Inter quod mare et gog et ma
gog fortissime gentes ab alexandro incluse dicuntur. partia ab oriente habet Indum fluvium a meridie mare indicum.
ad rubrum ab occidente mediam a septentrione yrcanum salum. Parti a syris originem extraxerunt fuerunt enim
exules quod parti sonat attica lingua et hec iuxta yrcania decem fines virtute occupaverunt. In ea est asyria media et
persia. Asyria. dicta ab asur ab oriente habet indiam medium tamen est gangandes per. a meridie tangit mediam
ab occidente tygrim et mesopotamiam. a septentrione caucasum. Est alius caucasus. qui a caspio mari
orientis attollitur et per aquilonem vergens pene usque ad europam porrigitur et pro gentium ac lin
guarum varietate in diversis sui partibus diversimode nominatur. v. enim in oriente excelsior est
propter nivium candorem caucasus dicitur. alibi dicitur infates. alibi sapedon alibi porte
caspie. alibi mitanus alibi corasicus alibi sitiacus alibi ceraunia alibi montes ar
menie et indie. hunc inhabitant amacones masagete colchi sarmate. In iudice cau
casi superavit alexander drancas evergetas parimas parpamenos adriaspios cete
rosque populos. Media dicta a medo rege ab oriente caspios videlicet a meridie
persidam. ab occasu transversa partie regna amplectitur. a septentrione arme
nia circundatur. In una parte habitant saraceni. in alia vocati corosinus. est
autem duplex media maior et minor. persia dicta a perseo rege qui ex
grecia transiit. ab oriente tendit usque ad Indos. a meridie carmaniam que
persidi annectitur ubi susa oppidum ab occasu rubrum mare ab aqui
lone mediam. sichia. sciticus et gotia a magog filio Japhet cogno
minata a dextra orientis parte ubi occeanus sericus est extendit
usque ad mare caspium quod est ad occasum ad meridiem vero
dehinc usque ad caucasum et subiacet ei yrcania. yrcania
ab oriente mare caspium a meridie armenia ab occidente
ybernia postquam capadocia ab aquilone albania ab yrca
nia silva nomen accepit. albania. sic dicta eo quod albo na
scantur crine. Hec ab oriente sub mari caspio consurgens per
oram occeani septentrionalis ad meotidas paludes per de
serta extenditur.

Europa. a fluvio tanay usque ad fines yspanie porrigi
tur et in Insula gades finitur continens infrascriptas
provincias. Sitia inferior. a meotidis paludibus usque
germaniam porrigitur et propter barbaros inhabitantes
barbarica dicitur. continet autem alaniam parte prima
que ad meotides paludes pertingit post hanc daciam
ubi et gotia. Germania. post sitiam ad orientem
habet danubium a meridie renum ab occiduo et septen
trione occeanum iuxta quod est germania superior sicut
circa renum inferior. Mesia ab oriente hostijs da
nubij iungitur a meridie macedonie ab occasu ystrie
a septentrione danubius dividens a barbarico ubi sunt
vandali rugi eruli turcilingi vinuli qui postea ytaliam
possederunt et longobardi dicti sunt. scitobini scori
gimaurngi golandi asipiti bulgares qui hunni et ca
gora dicti sunt. Rugilandi gepidi, sarmate sueui pa
noni qui usque ytaliam extenderunt. saxones norici qui
et baioarij qui ab oriente habent panoniam ab occidente sueviam
a meridie ytaliam. ab aquilone danubium. Tracia. ab oriente
habet constantinopolim. a meridie egeum mare ab occasu egeum
mare a sept. ystria obtenditur. grecia olim dicta tota cetus sive ito
norum et ylicum generaliter omnis grecia est vij. infrascrip
tas continet provincias. dalmatia. est ab occidente huius ab ori
ente macedoniam a meridie mare adriaticum ab occasu y
striam a septentrione messiam. Epyrus nominata ab epyro
achilis filio cuius pars caonia que ante molosia dicta est.
Elados. nominata a rege elana media inter macedoniam et achaiam
archadie a septentrione iungitur eladis due sunt provincie boetia
in qua tebe et pelopensie. Thesalia a meridie macedonie coniuncta
est cui pyria a tergo est in ea mons pernausus constans apolini consecratus.
Macedonia ab oriente coheret egeo mari a meridie achaie ab occasu
dalmatie a septentrione messie. Achaia. pene insula est a septentri
one tantum macedonie iungitur. ab oriente habet mirtenum mare ab euro
creticum ab meridie Jonium ab africo et occasu cassiopas insulas huius caput co
rinthus archadia sinus achaie in Jonium et egeum mare exposita. ipsa est sycionia
panonia alpibus apenninis ab ytalia secernitur ab oriente habens messiam ab euro
ystriam ab africo alpes ab occasu galliam belgicam. Lacedemonia habet ab oriente
messiam ab euro ystriam ab africo montes appenninos ab occasu galliam belgicam a
septentrione flumen quod galliam et germaniam dividit. ytalia patet. Germania
ab oriente panonie ungarie et boemie iungitur a meridie ytalie. Gallia. ab oriente habet
alpium iuga a meridie prerupta pyrenei ab occasu occeanum a septentrione renum et ger
maniam. yspania patet.
Vero melius. Jungitur ducatum athenarum quod aliter fere a tribus partibus maribus circumdatur. s. ab oriente meridie et plures sunt insule et occasu
ex septentrione recipies per magna parte iacet sed mare.

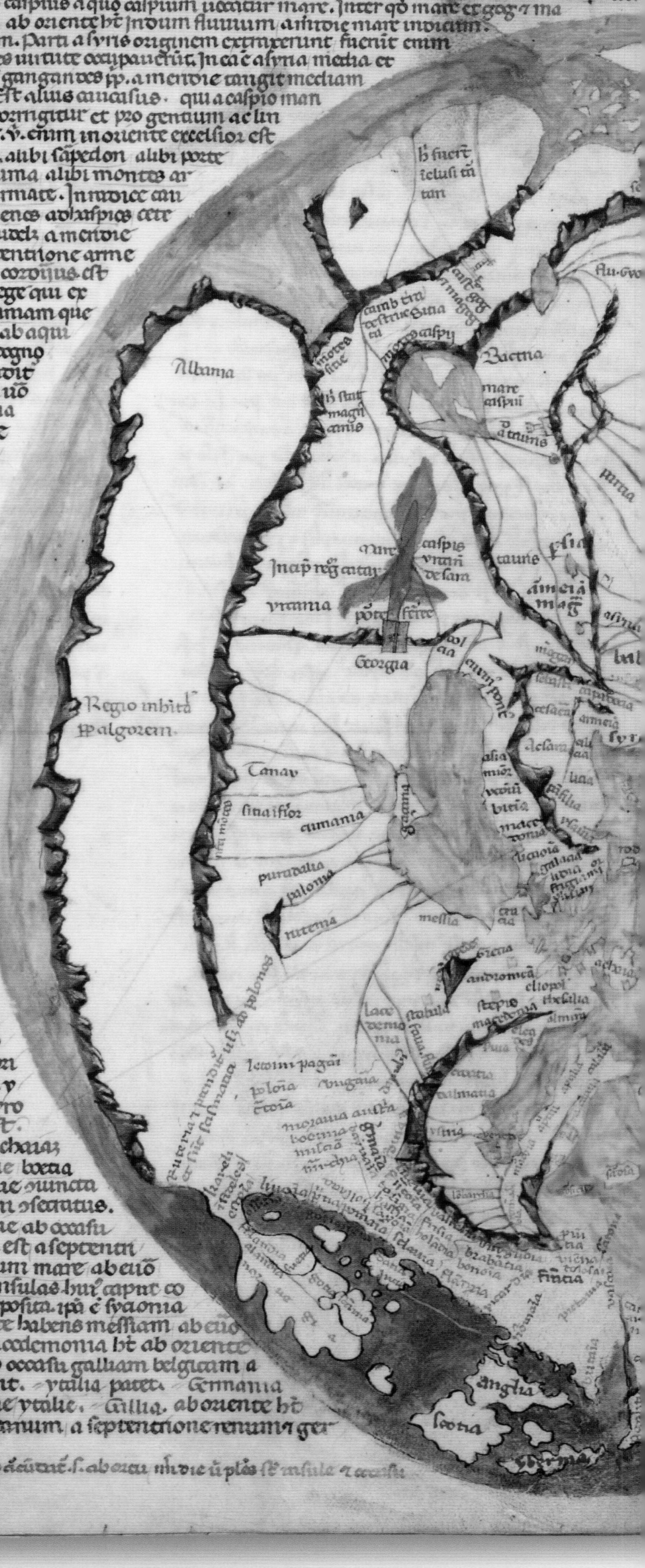

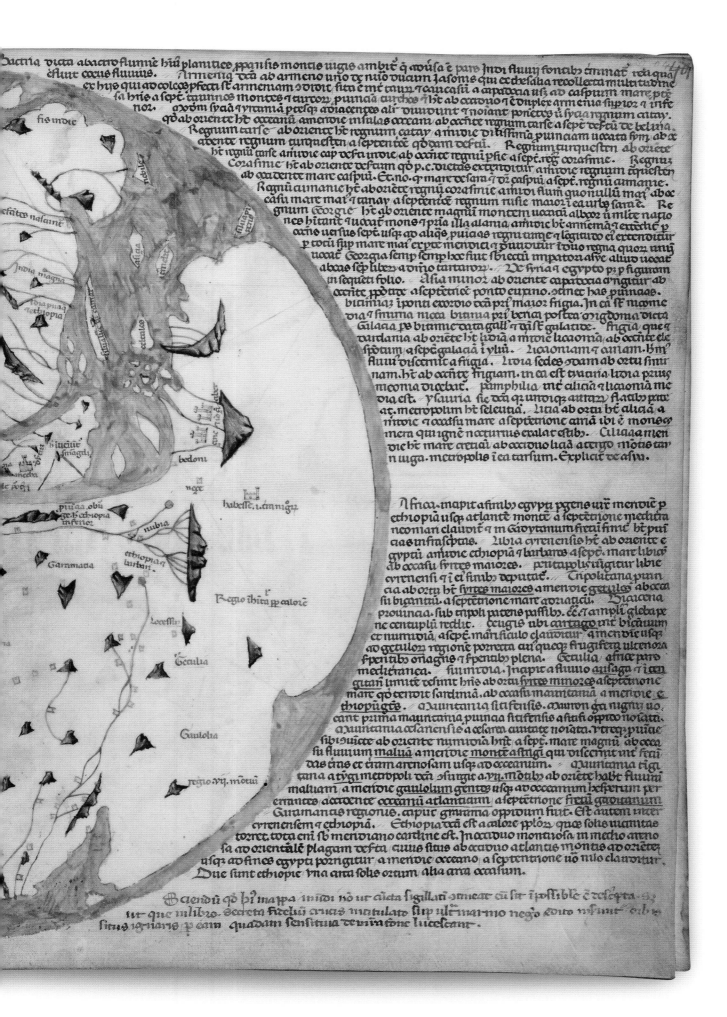

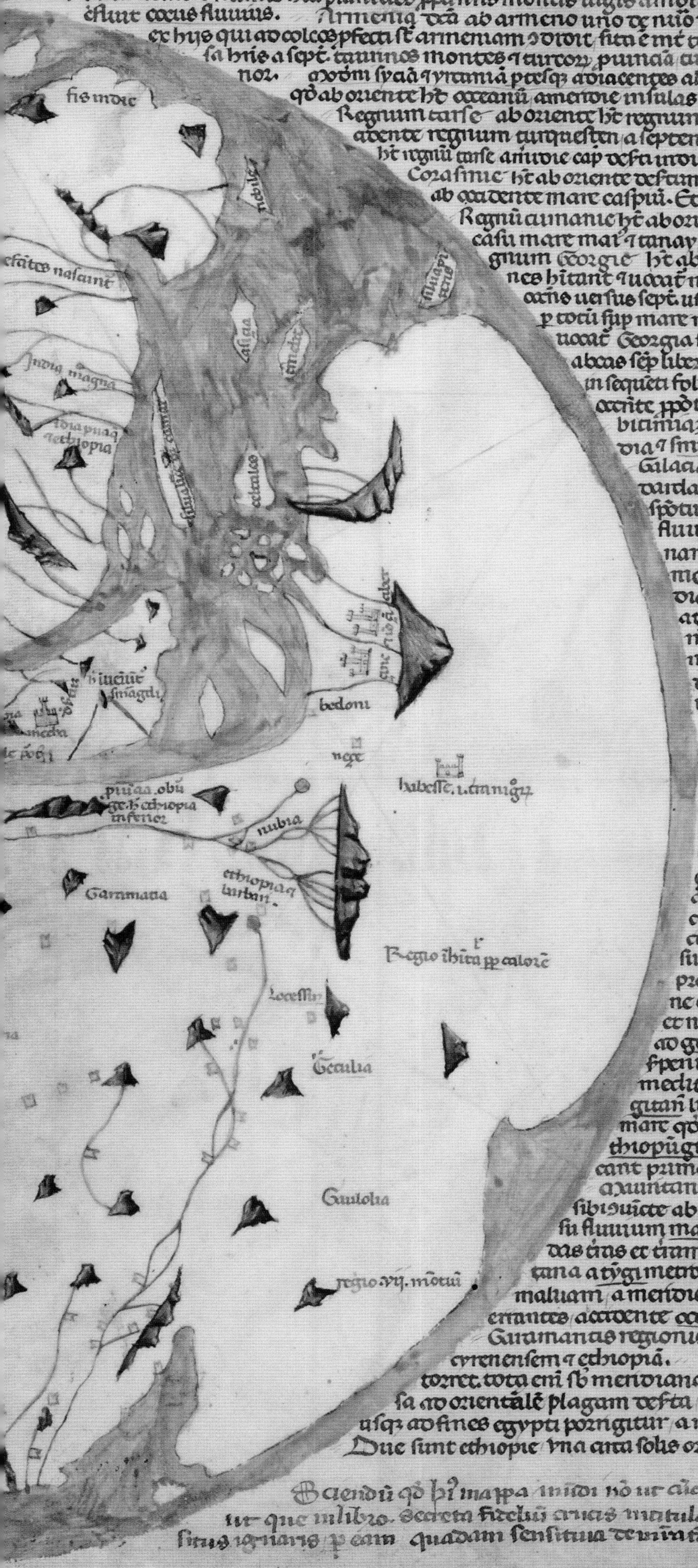

Bactria dicta a bactro flumine huius planities ppansis montis iugis ambit q adusa e pars Indi fluuii fontibus terminat reliqua efluit coctus fluuius. Armenia dcā ab armeno uno de ducibus iasonis qui ex thesalia recollecta multitudine ex hys qui ad colcos pfecti sunt armeniam condidit. sita e int taurum et caucasum a capadocia usque ad caspium mare ptē sa hns a sept. taurinos montes et turcorum puinciā turchos. hēt ab occiduo et ē duplex armenia supior et inferior. Medorum syriā et yrcaniā pterq3 adiacentes aliquas diuidunt et noiant ponetes q̃ syria regnum catay. qd ab oriente hēt occeanū a meridie insulas occeani ab occidente regnum tarse a sept. desertū de belina. Regnum tarse ab oriente hēt regnum catay a meridie ditissimā puinciam uocata sym ab occidente regnum turquesten a septentrione qd dam desertū. Regnum turquesten ab oriente hēt regnū tarse a meridie cap desti indie ab occidente regnū phie a sept. reg corasmie. Regnum Corasmie hēt ab oriente desertum qd p. c. dietas extenditur a meridie regnum turquesten ab occidente mare caspiū. Et no. q mare de sara et dr caspiū a sept. regnū cumanie. Regnū cumanie hēt ab oriente regnū corasmie a meridie flum quo nullū mar ab occasu mare mai et tanay a septentrione regnum rusie maior i ea urbs sara ē. Regnum Georgie hēt ab oriente magnū montem uocatū alborz ubi multe nationes hitant et uocat mons et pria illa alania a meridie hēt armeniā et extendit p occidens uersus sept. usq3 ad aliqs puincias regni turquie et longitudo ei extenditur p totū sup mare mai expte meridiei et diuiditur i duo regna quorum unum uocat Georgia semp semp hoc fuit sbiectū imperatori aliud uocat abcas sēp liberum a dominio tartarorum. De syria et egypto pz p figuram in sequenti folio. Asia minor ab oriente capadocia cingitur ab occidente ppontide a septentrione ponto euxino. continet has puincias. bitiniam i ponti exordio dcā prius maior frigia. In ea sunt nicomedia et smirna nicea bitinia prius berica postea migdonia dicta Galacia p9 bitinie data gallis et dicti sunt galatide. Frigia que et dardania ab oriente hēt lidiā a meridie licaoniā ab occidente ellespōtum a sept. galaciā i ylium. Licaoniam et ciliciam hermus fluuius discernit a frigia. Lidia sedes condam ab ortu simulnam hēt ab occidente frigiam. in ea est tyatira. lidia prius meonia dicebatur. Pamphilia int ciliciā et licaoniā media est. Ysauria sic dcā q3 undiq3 circa flatibus perciat. metropolim hēt seleuciā. Licia ab ortu hēt ciliciā a meridie et occasu mare a septentrione cariā ibi ē mons chimera qui ignē nocturnis exalat estibus. Cilicia a meridie hēt mare creticū ab occiduo liciā a tergo montis taurum iuga. metropolis i ea tarsum. Explicit de asia.

Africa. incipit a finibus egypti pgens iuxta meridie p ethiopiā usq3 atlantē montē a septentrione mediterraneo mari clauditur et in Gaditanum fretū finitur hēt puincias infrascriptas. Libia cyrenensis hēt ab oriente egyptū a meridie ethiopiā et barbaros a sept. mare libicū ab occasu syrtes maiores. Pentapolis iungitur libie cyrenensi et i ei finibus deputatur. Tripolitana puincia ab ortu hēt syrtes maiores a meridie getulos ab occasu bizantiū. a septentrione mare adriaticū. Bizacena prouincia. sub tripoli patens passibus. ccc et amplius gleba pre ne centuplū reddit. Zeugis ubi cartago int bizantium et numidiā. a sept. mari siculo clauditur a meridie usq3 ad getulorum regionē porrecta cui queq3 frugifera ulteriora ferentibus onagris et serpentibus plena. Getulia affrice pars mediterranea. Numidia. Incipit a fluuio ausaga et i zeugitani limite desinit hns ab ortu syrtes minores a septentrione mare qd tendit sardiniā. ab occasu mauritaniā a meridie ethiopū gentes. Mauritania sitifensis. Mauron greci nigrum uocant prima mauritania puincia sitifensis a sitifi oppido noiata. Mauritania cesariensis a cesarea ciuitate noiata. Utreq3 puincie sibi coniuncte ab oriente numidiā hnt a sept. mare magnū ab occasu fluuium maluā a meridie montē astrigi qui discernit int fecundas terras et terram arenosam usq3 ad occeanum. Mauritania tingitana a tyngi metropoli dcā consurgit a. vij. montibus ab oriente habet fluuium maluam a meridie gaulolum gentes usq3 ad occeanum hesperium pererrantes a occidente occeanū atlanticum a septentrione fretū gaditanum Garamantas regionis. caput garamantia oppidum fuit. Est autem inter cyrenensem et ethiopiā. Ethiopia dcā est a calore populorum quos solis uicinitas torret. tota enim sub meridiano cardine est. In occiduo montuosa in medio arenosa ad orientalē plagam deserta cuius situs ab occiduo atlantis montis ad orientem usq3 ad fines egypti porrigitur a meridie occeano a septentrione nilo clauditur. Due sunt ethiopie una circa solis ortum alia circa occasum.

Sciendū qd hec mappa mundi no ut cuncta sigillatim contineat cū sit impossibile e descripta sz ut que in libro. secreta fidelium crucis intitulato sup ultramarino negocio edito inserunt oculis situs ignaris p eam quadam sensitiua demonstratione lucescant.

not translated into Latin at that time, so it would have been of limited use for an increasingly monolingual administration. Moreover, the era of the multifaith experiment of Norman Sicily was coming to an end. Over the next few decades, the Greek-speaking and Arabic-speaking members of the elite lost their status in the court. In the 1220s, following a major rebellion, the Muslim communities of Sicily were expelled to Lucera on the Italian mainland, effectively ending the Muslim and Arab presence on the island for the remainder of the Middle Ages.

Yet some version of the circular world map must have been circulating in Italy, which helped medieval Europeans to conceive of the geography of the world in new ways. In 1321 Marino Sanudo, a Venetian merchant campaigning for the revival of the Crusades, presented to the pope a book called *Secrets for True Crusaders*. This treatise was part biblical history, part history of the Crusades, and it contained a practical section on the geography of the Holy Land and on strategies for reclaiming it. It also included a world map (previous spread) prepared by the Genoese map-maker Pietro Vesconte. Despite the context of the treatise, and perhaps because of it, Vesconte offered Europeans a far less theological view of the world than the traditional *mappae mundi* tradition, which often depicted paradise in the east and Jerusalem at the centre of the world. Vesconte's map is still oriented to the east, but it is devoid of religious content. The precision in the coastlines of the Mediterranean is primarily a result of new techniques of nautical charting (which will be discussed in Chapter 5). But the influence of al-Idrīsī's circular world map is evident in the shape of the continents, in the introduction of a western Nile that flows to the Atlantic, and in a few labels in the northern regions, on the left-hand side, that repeat al-Idrīsī's Arabic terms for the unpopulated areas of Siberia.

In the Muslim world, al-Idrīsī was best known among North African scholars. There are ten manuscript copies of the *Entertainment*, the earliest from around 1300, the latest from the end of the sixteenth century, and all appear to have been made in the Maghreb or in Muslim Spain. But the intellectual legacy of al-Idrīsī goes beyond the surviving copies of his treatise. No other medieval geographer had such a universalist approach to human geography, and no other medieval cartographer made such an ambitious attempt to grasp the interconnectedness of different regions of the known world. So when the North African historian Ibn Khaldūn began to write a universal history of the world, he took his geographical introduction largely from al-Idrīsī's work. Ibn Khaldūn even decided to preface his most famous work, his universalist philosophy of history known as the

(previous spread) A world map prepared by the Genoese map-maker Pietro Vesconte, 1321. Bodleian Library, University of Oxford, MS. Tanner 190, fols 203v–204r.

Muqaddimah, with al-Idrīsī's circular world map. It is included in a manuscript probably made under the supervision of Ibn Khaldūn himself in 1401. As Tarek Kahlaoui has suggested, al-Idrīsī's visual and textual conception of physical and human geography facilitated Ibn Khaldūn's development of his universal rules of human history.[19]

Al-Idrīsī also left us some unique observations on the purpose and power of maps, as well as their limitations. He writes that he prepared each of the seventy sectional maps so:

> That the one who observes it can see that which is hidden from his sight, or not known to him, or would not be able to reach himself due to the difficulty of the roads and the differences between nations. But through observation of these maps he is able to grasp this knowledge accurately.[20]

A map allows viewers to see with their own eyes, and therefore to accurately grasp geographical knowledge. It can show them lands they will never visit, information that they do not know, or reveal to them, through the power of the visual image, what is otherwise hidden from their sight – even in lands that are familiar. In al-Idrīsī's cartographical imagination, maps still need to be explained and expanded, and the *Entertainment* is a book full of textual information about the kingdoms and peoples of the world. But a true observation of the world can come only through a good map.

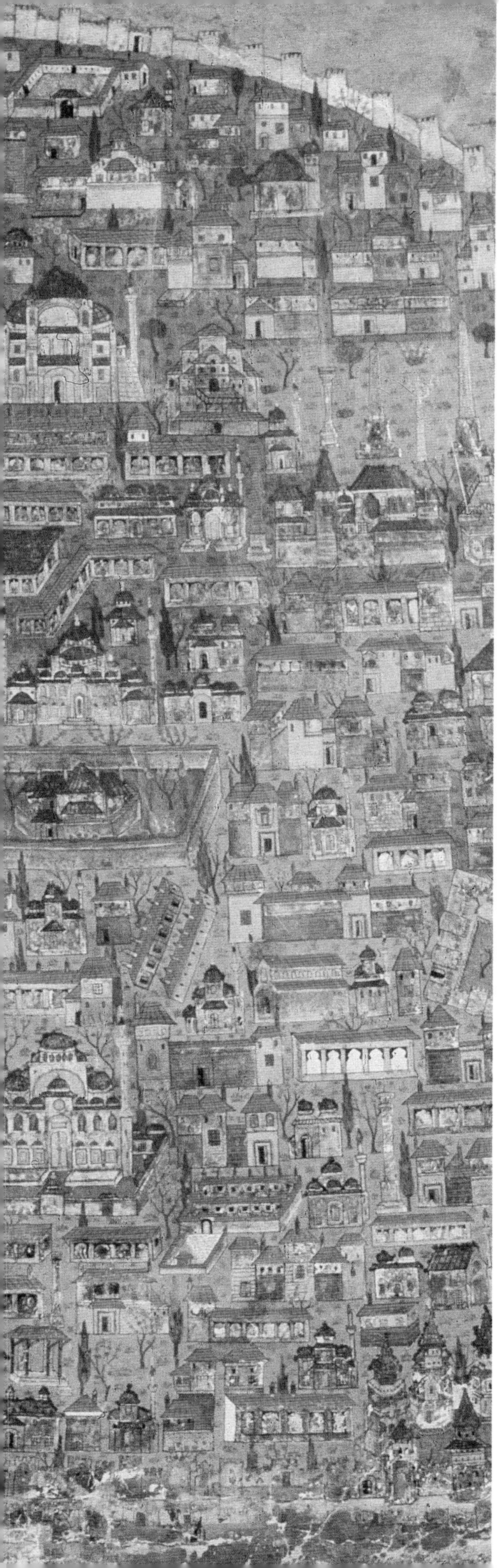

CHAPTER 5

The Expanding Horizons of an Ottoman Admiral

IN THE SPRING OF 1554 two light galleys entered the Ottoman port of Suez, carrying Pīrī Reʾīs, the experienced admiral of the Ottoman Indian Ocean fleet. It was a sorry sight. Pīrī had left Suez in the previous year with twenty-four galleys and four bargias (heavy warships) under his command, set on dislodging the Portuguese from the strategic fort of Hormuz, the key to control of the Persian Gulf. After stopping in Aden, Pīrī captured the port of Muscat, and then continued to besiege the undermanned Portuguese garrison in Hormuz. But, upon receiving news of Portuguese reinforcements coming from Goa in India, he decided to withdraw his fleet to the safety of Ottoman-held Basra. Then, in an act of personal bravery, he took his fastest galleys and made his way through the Portuguese naval blockade back to Egypt, leaving the main body of the fleet stranded in Basra. In Egypt, however, his actions were seen as treachery or as cowardice, and the failure to capture Hormuz was perceived as a crucial defeat in the Ottoman–Portuguese battle for control of the Indian Ocean. Pīrī was placed under arrest and, following a direct order from Sultan Suleiman, was executed in Cairo, probably in the summer of the same year.

It was a tragic end to the long life of one of the finest sea captains of the age of discovery and, without doubt, the greatest Ottoman map-maker of all time. Pīrī Reʾīs was of the generation of Ferdinand Magellan, Vasco da Gama, John Cabot and Hayreddin Barbarossa. Like Francis Drake, Pīrī was both a pirate and a loyal admiral in the service of his country, a man of adventure who made the most of the infinite opportunities the sea suddenly opened up in the sixteenth century. But, unlike most other famed sea captains of his age, Pīrī left us an incredible cartographic corpus, including one of the earliest world maps to show the Americas and detailed charts of Mediterranean islands and coasts in his *Book on Seafaring*. His maps were heavily indebted to Portuguese informants, Catalan charts and Italian modes of visual expression. But they were also highly distinctive, unlike any maps produced in Europe at the time. Like all cartographers, Pīrī hoped his maps would secure him prestige and patronage. But, at a deeper level, he used maps to express wonder and awe at the discovery of new lands and the spirit of unshackled maritime adventure. His maps and the accompanying texts can be read as an ode to the sea where he spent his life.[1]

Ever since the Crusades, Muslim power had been trailing that of the Europeans in the battle over hegemony in the Mediterranean. In the eleventh century, the author of the *Book of Curiosities* still drew maps that represented the maritime ambitions of the Fatimid caliphs based in Cairo. But, with the onset of the

Crusades, Italian navies came to dominate the eastern Mediterranean. By the twelfth century, al-Idrīsī was producing his maps in the capital of Norman Sicily, which was never regained by the Muslims. For the remainder of the Middle Ages, the sultans of Cairo withdrew from the sea. They demolished many of the port installations in the eastern Mediterranean, such as the port of Tinnīs, which had been so beautifully drawn in the *Book of Curiosities*. Mediterranean trade was all channelled to the port of Alexandria, where Italian and Catalan merchants could be closely supervised.

European domination in the Mediterranean was consolidated through advancements in navigation brought about by the combined introduction of the magnetic compass and the portolan chart. The portolan chart, first attested around 1270 in Pisa, signalled a revolution in maritime and European map-making. Until then, maps made in Latin Europe were mostly diagrammatical sketches that accompanied historical and theological narratives. Portolan charts, however, were practical instruments of navigation, developed by sailors and captains who were using the magnetic compass to coast along the Mediterranean shores. The gradual accumulation of information on the direction of the coastlines, gleaned through the consistent recording of compass bearings, allowed map-makers in Genoa and Mallorca to produce charts that reproduced the shape of the Mediterranean roughly as we know it today. These charts were then taken to sea, where they allowed pilots to travel away from the coast with unprecedented confidence.

Muslim mariners were quick to appreciate the value of the new portolan charts, and North African map-makers began to draw them up in Arabic. The first description of a portolan chart by an Arab scholar dates to the 1340s. Its author, al-ʿUmarī, called the new chart a *qunbās*, from *compasso* in Italian, and explained the function of the distinctive rhumb lines that allowed ships to follow a course indicated by the magnetic compass. Towards the end of the fourteenth century, the Tunisian scholar Ibn Khaldūn mentioned the use of *qunbās* charts by North African mariners. He also explained that these charts only covered the Mediterranean but not the Encompassing Sea to the west, to which sailors therefore did not dare to sail: were they to lose sight of the shore, they would not be able to find their way back.[2]

A handful of Arabic portolan charts from the Late Middle Ages survive, all produced in the ports of North Africa. A relatively late example, produced by a family workshop in the Tunisian city of Sfax in 1571–2, covers the central areas

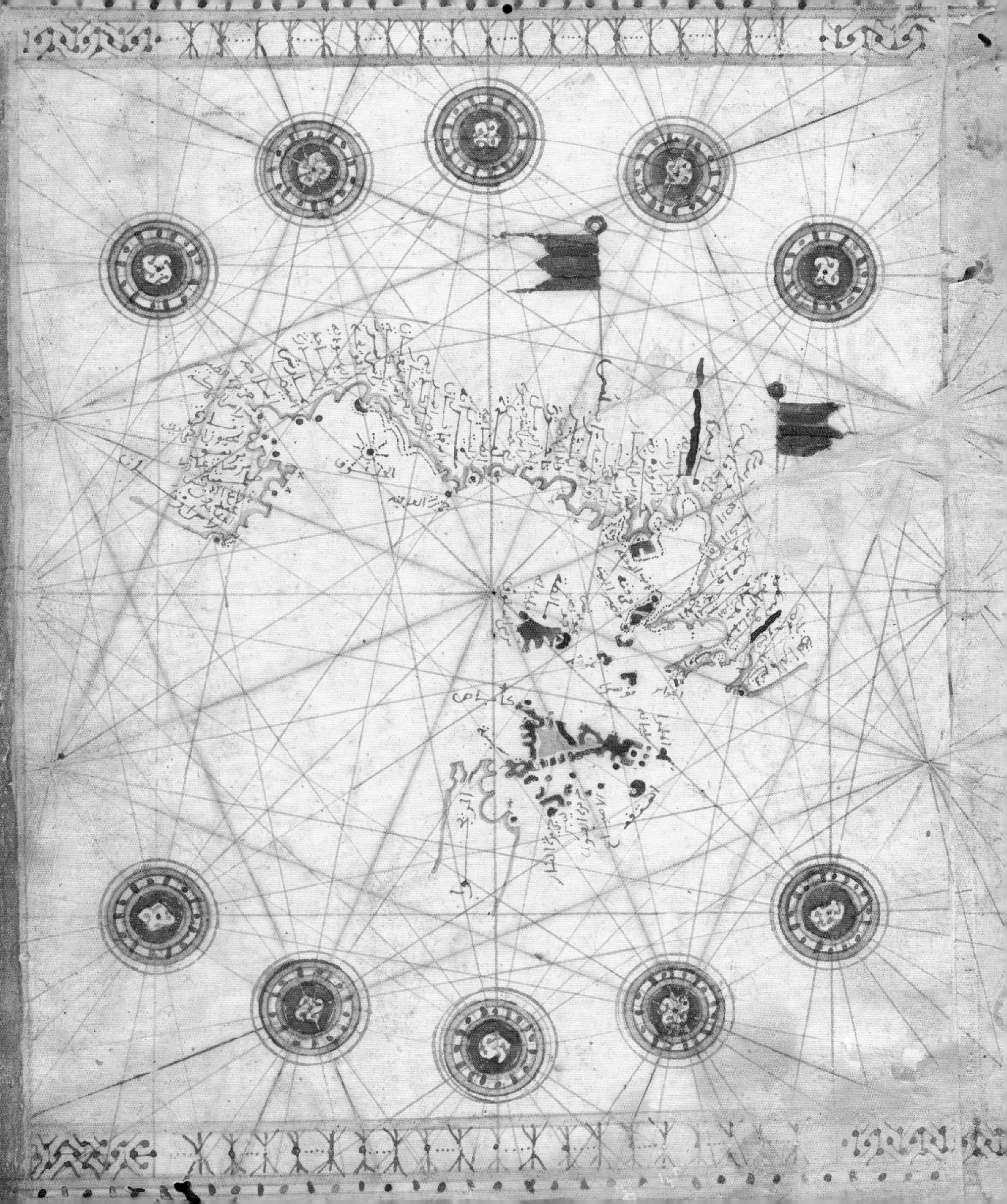

Arabic portolan chart from a family workshop in the Tunisian city of Sfax, 1571–2, showing the central areas of the Mediterranean. South is at the top. Bodleian Library, University of Oxford, MS. Marsh 294, fol. 6a.

of the Mediterranean (opposite). The chart is oriented to the south, with the oversized island of Malta near the intersection of the radiating rhumb lines. As in Italian and Catalan portolan charts of the same period, the place names are perpendicular to the coast, and islands are characteristically depicted in brilliant colours. The Arabic script is distinctively North African, and an inscription in another copy of this map mentions that it is made on the basis of a Mallorcan chart. Somewhat crude, and far less ornamental than the numerous Catalan portolan charts known to us, this derivative chart would have provided North African mariners with an approximation of the contours of the Mediterranean shores and the direction of sail away from the coast.

The Ottoman conquest of Constantinople in 1453 changed the balance of power in the Mediterranean, bringing European hegemony to an end. The Ottoman sultans found themselves ruling over a global empire, and heirs to Byzantium's maritime glory. As part of this new imperial outlook, Sultan Mehmed II assembled any map of the world he could get hold of, including copies of al-Istakhrī's maps (discussed in Chapter 2), maps based on Ptolemy produced by Greek scholars, and a couple of fifteenth-century North African portolan charts, which are still preserved in the Ottoman palaces of Istanbul. His successor, Bayezid II (1481–1512), actively pursued an ambitious expansion of the Ottoman navy by approaching Turkish corsairs known to be operating in the Mediterranean and offering them positions in the Ottoman navy. Co-opting the pirates paid off: within a couple of decades, Ottoman vessels were setting out to capture Venetian strongholds all along the Peloponnesus.

The World Map of Pīrī Reʾīs

Pīrī Reʾīs, born around 1470, was one of the pirates recruited in this way to the Ottoman navy. Since 1487 he had accompanied his uncle, Kemal Reʾīs, a successful pirate, along the North African coast. Kemal Reʾīs operated from a base in the inaccessible island of Jerba, off the coast of Tunisia, whence he travelled far and wide in the western Mediterranean. His main target was the rich pickings to be had from commercial vessels crossing the Tunisia–Sicily bottleneck in the central Mediterranean. He also raided the Balearic Islands and Corsica, raids in which the young Pīrī participated. In 1495 Kemal Reʾīs was summoned to Istanbul, and offered a commanding position in the Ottoman navy by Bayezid himself. Pīrī and his uncle were now based at the naval headquarters in Gallipoli. By 1499 Pīrī was a *reʾīs*, a captain, in command of his own ship, as part of an Ottoman force that captured

Lepanto from the Venetians. Kemal probably died in 1510 or 1511, when his ship was wrecked by a storm on his way to capture Rhodes, where the Knights of the Order of St John were to hold out against the Ottoman navy for another decade.

Beyond the horizons of Venetian–Ottoman squabbles in the Aegean, however, the world was expanding at astonishing pace. Pīrī was still operating as a pirate in the western Mediterranean when news arrived of Columbus's voyages and the discovery of the New World. By 1499 Vasco da Gama had circumnavigated Africa, and within a decade the Portuguese were becoming a threat to the Indian Ocean trade, a major source of revenue for several Muslim dynasties, most prominently the Mamluks of Cairo. Another Portuguese sailor, Pedro Álvares Cabral, encountered the coast of Brazil in 1500. This vast new geographical knowledge was almost immediately translated into maps, and the greatest and fastest advances were made by the Portuguese. As can be seen in the 1502 world map known as the Cantino planisphere, Portuguese map-makers extended and applied the rhumb lines and wind roses used in creating the portolan charts of the Mediterranean. Even when these were applied to the mapping of the world's oceans, the results were remarkably accurate.

Back in Gallipoli, Pīrī took notice. In the spring of 1513, only two decades after the voyages of Columbus, Pīrī completed a world map of his own, one of the most famous and intriguing maps in the history of cartography. Drawn on several pieces of parchment, this was a massive object, probably measuring 140 by 165 centimetres. Of this original, only the central and southern parts of the western third have survived on a fragment measuring 87 by 63 centimetres (opposite). The Atlantic is at the centre of the map. In the east, the fragment covers the Atlantic coasts of France, the Iberian Peninsula, North Africa and the Gulf of Guinea. In the west, the fragment shows the Caribbean, Cuba and the Bahamas. The Brazilian land mass is most prominent and easily recognizable. It is joined to a land mass to the south, presumably the Terra Australis, as was common in early modern European world maps.

Pīrī's world map is among the most valuable and well-accomplished works of the age of discovery. The technique used by Pīrī is clearly that of the men of the sea, not that of armchair landlubbers. The map has all the typical characteristics of portolan charts, such as the two thirty-two-point compass roses, the network of rhumb lines and the prominent length scales sloped around the tropics of Cancer and Capricorn. Navigational markings are visible along the shores: crosses for reefs, red for shallows and black dots for ledges. It clearly draws on Portuguese

The extant portion of the world map made by Pīrī Reʾīs in 1513, showing the Atlantic Ocean. Topkapi Palace Museum, Istanbul, R. 1633.

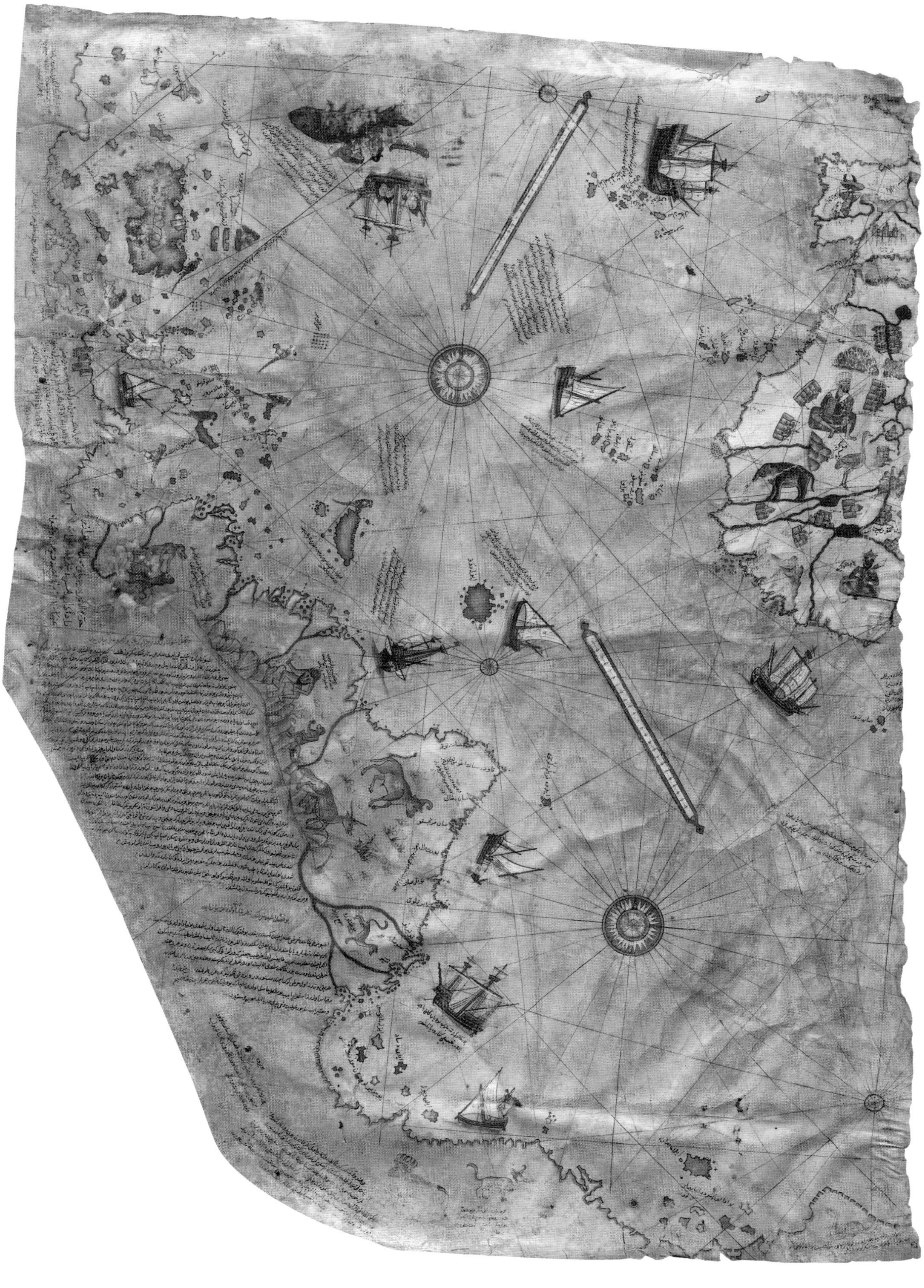

charts, such as the 1502 Cantino planisphere. But, compared to the Cantino map, it is also much richer. The surviving fragment alone has 114 place names, thirty-five longer inscriptions and a wide array of iconography.

Luckily for us, Pīrī's signature is found on the surviving fragment of the map, to the left of centre, written perpendicularly above the line of the equator and next to what looks like an image of a dog and a monkey holding hands. The signature (its technical name is a colophon) is in Arabic, while all the other inscriptions are in Ottoman Turkish. It is also in a different hand from that of most of the other labels on the map, which suggests that Pīrī employed professional chart-makers as assistants. But in the label Pīrī claims sole credit, stating that 'Pīr, son of Hacı Mehmed, also known as the nephew of Kemal Reʾīs, made this map in Gallipoli, on the month of Muḥarram of the year 919 [9 March–7 April 1513].'[3]

The long, scroll-like inscription that takes up most of the Brazilian land mass just below the signature narrates the discovery of the New World by Columbus. According to Pīrī, Kolomb of Genoa found an ancient book with information about these lands and their riches, and attempted to convince the kings of Europe to fund him. After receiving the support of the king of Spain, he sailed in the year 896 of the Islamic calendar (1490–1). He then encountered the shores of the lands Pīrī calls Antilla, where he met a range of indigenous people, many of them cannibals. The native peoples had also much gold and pearls, which they happily exchanged for glass beads. Pīrī concludes by noting that the Spanish king wants to convert the natives to Christianity.[4] Another inscription, just below the northern wind rose, indicates the meridian that marked the boundary between Spanish and Portuguese territories in the Americas, as agreed in the Treaty of Tordesillas in 1494.

The shorter inscription below sets out the sources Pīrī used for constructing the map; it has bewildered historians of cartography and exploration for over a century.[5] First, Pīrī is keen to emphasize that this is an original map: 'No one in this age possesses a map such as this one.' But he acknowledges that it is an act of synthesis. He consulted twenty charts and *papamondular*, a Turkish rendering of *mappae mundi*, and used nine maps of the inhabited quarter of the world made at the time of Alexander the Great, which the Arabs called the Jaʿfariyya (undoubtedly Ptolemaic maps in the tradition of mathematical geography). Additionally, he had before him four recently made Portuguese maps describing the seas of Sindh, India and China, and finally his source for the western lands –

the New World – is a map made by Columbus himself, which Pīrī combined with other maps and set at the same scale.

Did Pīrī really have in his possession a map made by Columbus? That seems unlikely, although his depiction of the Caribbean is particularly indebted to Columbus, with Cuba shown as a wedge of land jutting from the mainland and inclining southward, just as Columbus believed it to be. What does seem likely is that Pīrī had a Spanish informant. Elsewhere he mentions that his uncle Kemal Reʾīs had a Spanish slave (perhaps servant) who had accompanied Columbus on three voyages and knew much about the New World. Kemal must have accumulated quite a few Spanish captives during his raids in the western Mediterranean, and it is possible that one of them had gone to the New World, or at least claimed that he had in order to gain the respect of his captors.

Pīrī was keen not only to chart the seas of the world, but also to tell the story of their exploration. Visually, the iconography on Pīrī's world map is striking: it is not only richer than European world maps of the time but also marked a complete break from medieval Islamic map-making. There are some fifty-eight images on this fragment of people, beasts, vessels, mountains and plants. The rulers of Guinea, Marrakech and Portugal are individually shown, each with distinctive regalia and ethnic markers. An elephant dominates the African interior. The Americas, in particular, are replete with wonders. There are oxen with six horns, also shown in Portuguese maps, as well as monstrous snakes in the Terra Australis at the bottom of the map. A headless Blemmye is shown with hairy arms and a beard, a mouth on its chest and eyes on its shoulders. A monkey is depicted joining hands with a baboon-like animal that has the face of a dog. The most frequent image is of parrots: there are twelve different varieties of them on the map, all found on islands and all, according to the inscriptions, edible.

It is, however, the sailing vessels that are depicted with the utmost care, revealing the maritime sensibilities of a man of the sea. There are ten ships shown on the Atlantic, from triple-sailed galleasses to single-sailed caravels, their type usually easily identifiable through details in the image and the accompanying inscriptions. Each of the ships tells a story of a discovery or an adventure at sea. In the northern Atlantic the three-sailed ship is a Genoese galleass that has drifted off course en route from Flandres and landed in the Azores. The elaborate image in the north-west of the Atlantic illustrates the journey of St Brendan the priest, whose crew lit a fire on the back of a whale, mistaking it for an island (overleaf). Pīrī specifically points out that this information is not taken from the Portuguese

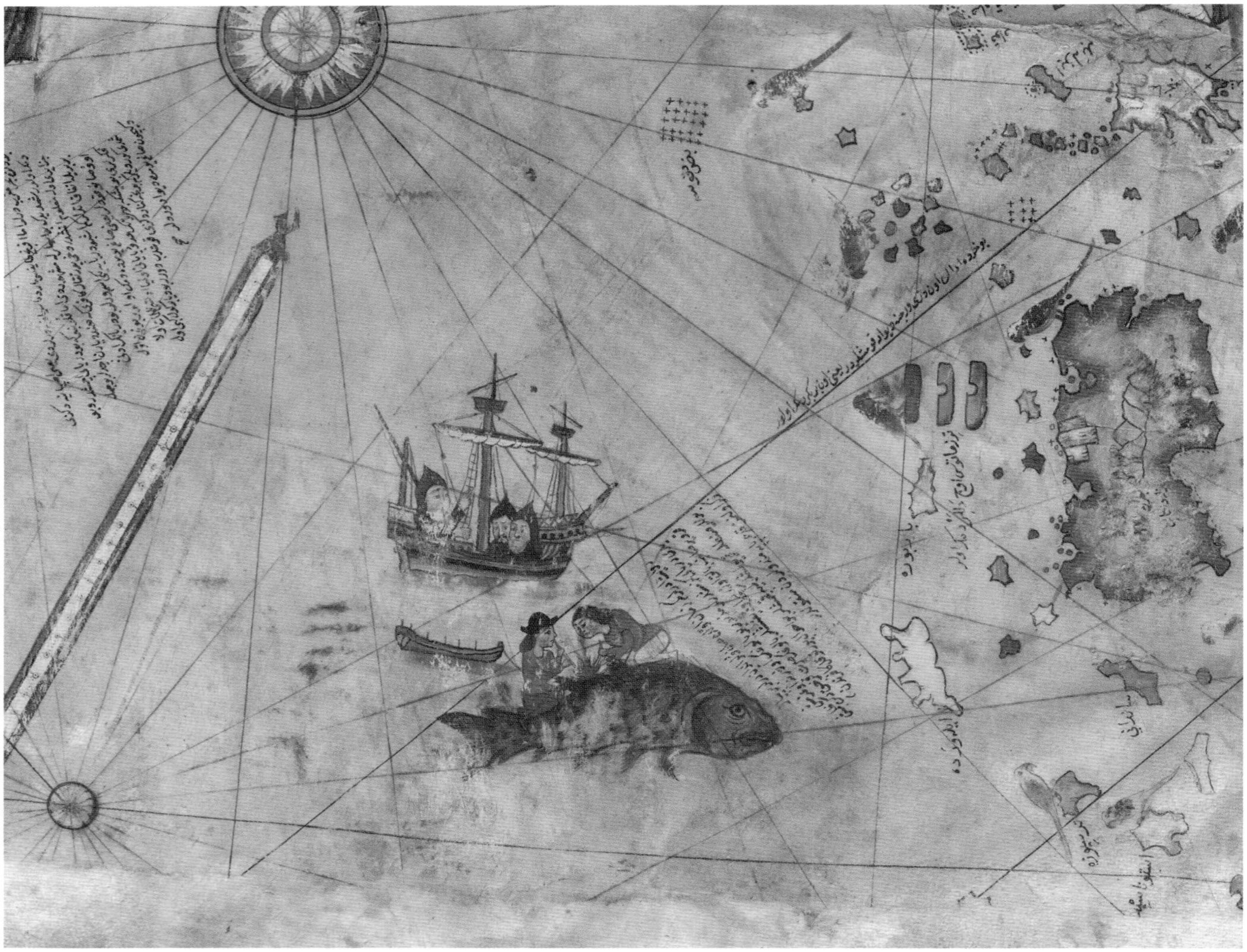

St Brendan riding on the back of a whale. Detail from the world map of Pīrī Re'īs, 1513. Topkapi Palace Museum, Istanbul, R. 1633.

charts but from an ancient *mappamondo*. The vessel shown off the west coast of Africa is apparently one of four vessels – others must have been drawn on the missing parts of the map – that are going to circumnavigate Africa under the command of Vasco da Gama.

After completing his world map in 1513, Pīrī waited for the right moment to present it to the new sultan, Selim, who had acceded to the throne in 1512. His chance came in 1517, following Selim's swift defeat of the Mamluk empire and the addition of Egypt and Syria to the Ottoman domains. In his *Book on Seafaring*, Pīrī tells us that he presented his world map to Selim in Cairo, shortly after it was (following spread left) The island of Cephalonia from the *Book on Seafaring* by Pīrī Re'īs, copied 1587. Bodleian Library, University of Oxford, MS. d'Orville 543, fol. 50.

(following spread right) The island of Rhodes from the *Book on Seafaring* by Pīrī Re'īs, copied 1587. Bodleian Library, University of Oxford, MS. d'Orville 543, fol. 31a.

taken by the Ottomans. For the Ottomans, the conquest of Egypt opened up new opportunities in the Red Sea and the Indian Ocean, and perhaps the depiction of these areas in Pīrī's world map was considered to be of strategic value. Someone at the Ottoman court then tore the map up, and only the Atlantic portion was placed in the palace libraries, where it remained unnoticed for four centuries. It was rediscovered only in 1929, when scholars began to mine the Topkapı Palace libraries in Istanbul following the abolition of the office of the Ottoman caliph by Republican Turkey.[6]

The *Book on Seafaring*

After his world map, Pīrī's great achievement is the *Book on Seafaring* (*Kitāb-i Baḥriyye*), whose first version was completed in 1521. The *Book on Seafaring* is a volume of sailing directions in the Mediterranean based on Pīrī's own experiences at sea. It is divided into 130 chapters, with each chapter devoted to a region or port and accompanied by a chart. The chapters start with the Aegean, and then move in counterclockwise direction along the Mediterranean coast, concluding with a set of Aegean Islands that were missed out earlier, and then with islands in the Sea of Marmara, near the Ottoman capital. The principal aim of the text, written in the second person, is to instruct skippers on short voyages from one safe haven to another. Each chapter describes landmarks, harbours, hazards and approaches to harbours. Much of the focus of the text is on protection from the prevailing winds, which posed a particular risk to the long and narrow galleys of the Ottoman navy.

Instead of describing the world on a large scale, Pīrī's maps are now precise and practical. The exclusive focus is on the coasts of the Mediterranean and, above all, its islands. The chart of the island of Cephalonia, the largest of the Ionian Islands in the Adriatic (p. 136) is typical of Pīrī's approach. In this copy of the *Book on Seafaring*, made in 1587 and now held in the Bodleian Library, the presentation is simple. This is a chart for mariners. Notice the use of bright colours, which would have helped navigators to distinguish between the different Ionian Islands and the mainland. There are indications of hills and of the main town on the island. Above all, the contours of the coastline of the island are shown with a level of precision not seen before in Islamic cartography, clearly based on compass readings and accompanied by a wind rose.

The compass was essential for the type of maps Pīrī was producing, as it had been for the makers of European and North African portolan charts in the

بر میل مقداری کشیشلمه طرفنده پشقارده‌دن غربو ایو لیماندر اول لیمانک اغزندن دمور قورسه قورلر
اون بش قولاجده‌در واکر ایچرو کیررلر ایسه یدی قولاجده‌در
ولیکن بیوک کمیلر ایکی جانبه پلمار باغلمق کرک طار یردر
وبعده غربو لیمانک مقابلسنده اولان
ایا ماوری بورنی اون میلدر
والسلام

علی التمام

اشکال جزیره

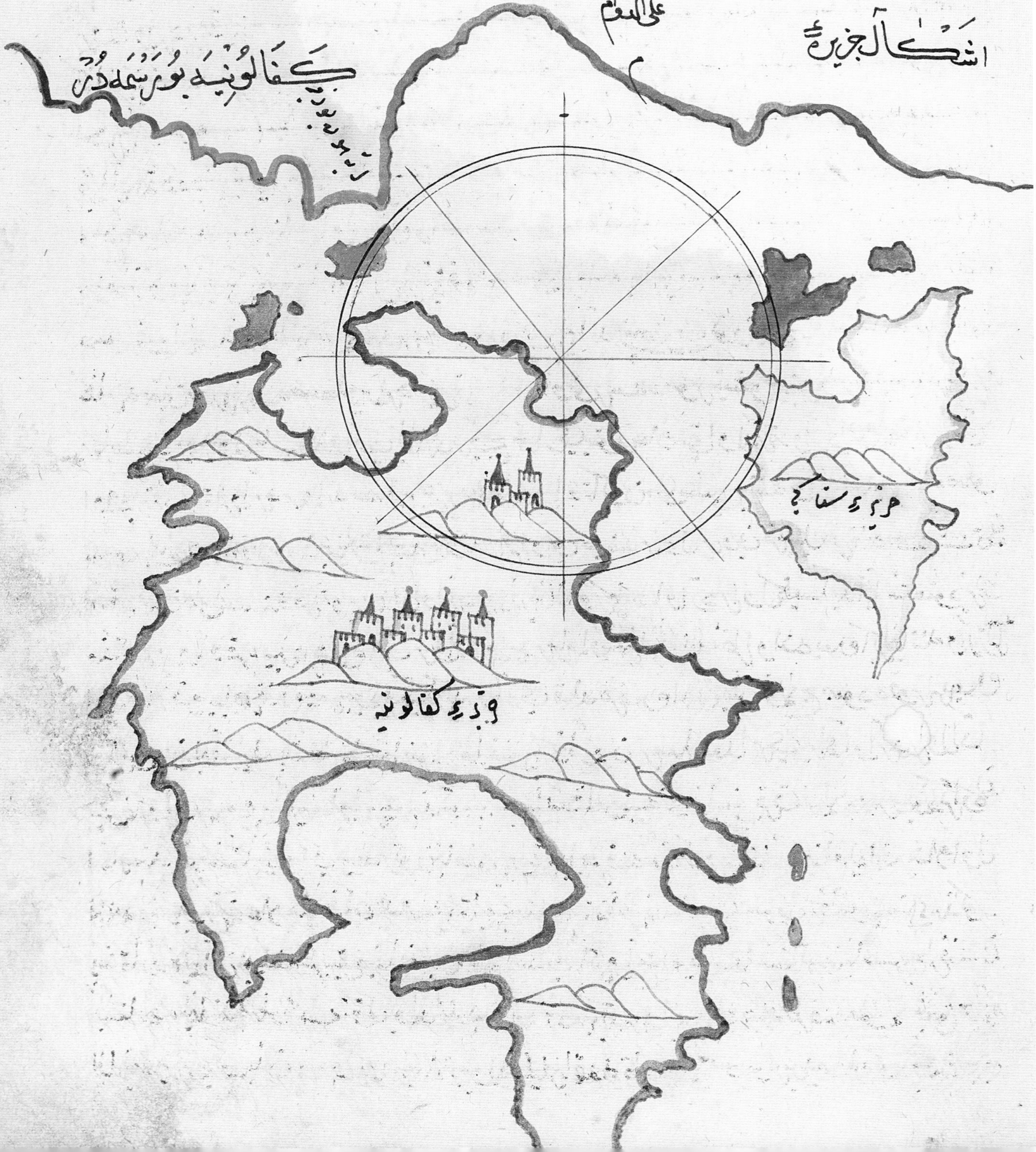

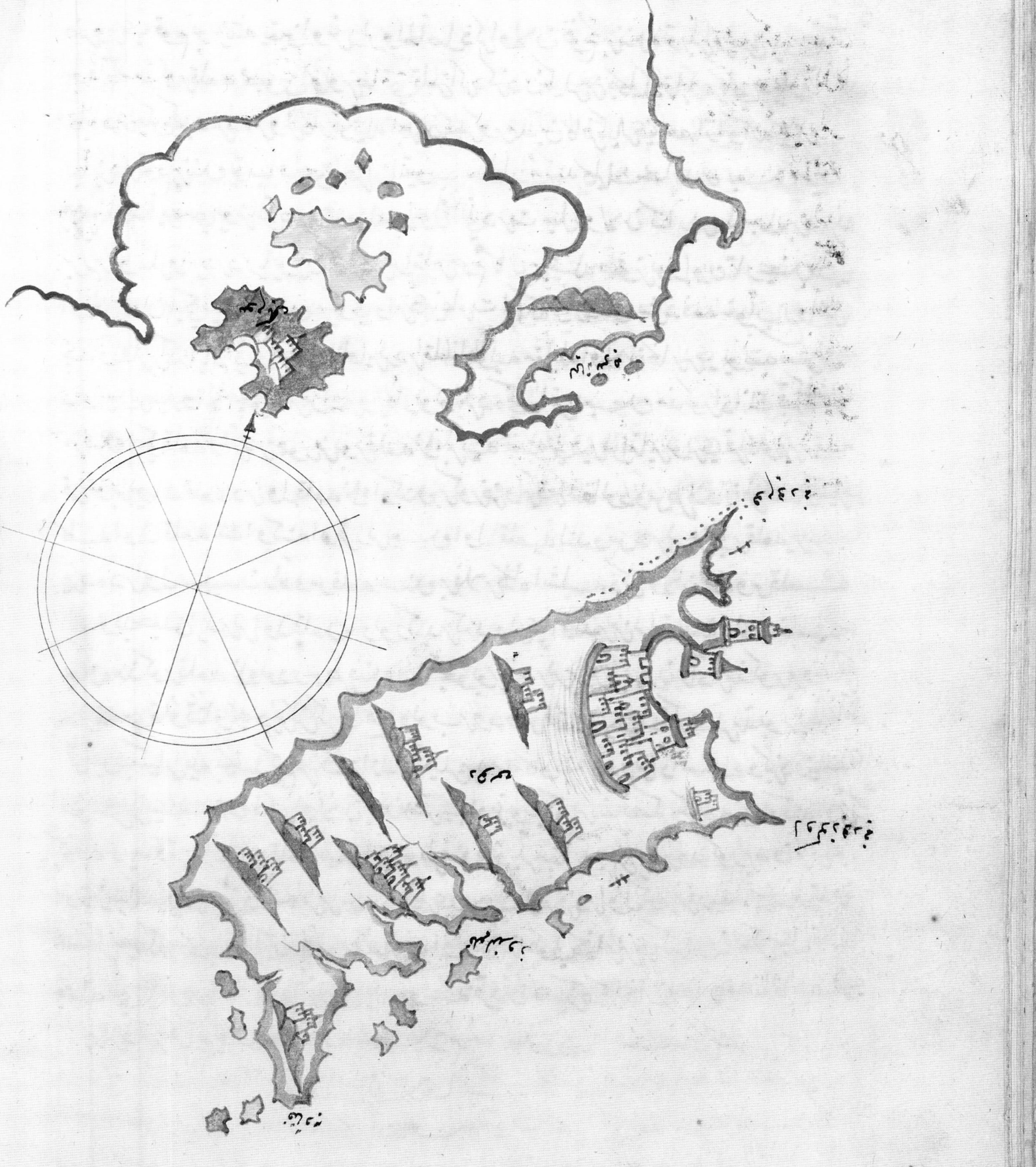

The island of Malta from the *Book on Seafaring* by Pīrī Re'īs, copied 1587. Bodleian Library, University of Oxford, MS. d'Orville 543, fol. 81b.

preceding centuries. In a versified introduction to the second version of the *Book on Seafaring*, Pīrī gives this description of the compass: '[The compass] is a round box, sealed in glass, containing a piece of paper with thirty-two corners. Its maker has placed this paper on a shaft of steel and having marked it permits it to rotate like the world. When it turns and comes to a halt, you may be sure that it points to the North Star.'[7] The charts, he then explains, are produced by the compass bearings, not by coordinates calculated by mathematicians. Coordinates are of no use at sea, and astrolabes, so useful for telling the time on land, are also of no value. For a sea chart, one must rely on the lines drawn on the basis of the wind roses and the compass, lines that bring together on the map localities that are not mutually visible in reality.

Pīrī was working within a tradition that had so far been dominated by European map-makers, but he was carving his own original path. The portolan charts produced in Europe or in North Africa usually covered the Mediterranean as a whole, or, as we have seen in the Sfaksī chart, wide sectors of it. Pīrī, for his part, chose to divide his text and his charts into much smaller units. In this he was influenced by a new and popular Italian genre of maps of islands, known as the *isolarii*, first introduced by Cristoforo Buondelmonti in 1420.[8] Like Buondelmonti and his successors, Pīrī decribes each island using image and text, but he expanded the pattern to the coasts of the mainland and directed his text clearly at mariners, specifically Turkish mariners. The text is written in Ottoman Turkish, not Arabic, and would have provided the Turkish-speaking pilots of the Ottoman navy with a practical guide to the entire waters of the Mediterranean.

The *Book on Seafaring* aided the Ottoman piecemeal expansion in the Mediterranean. The Ottomans advanced port by port and island by island, not in great pitched naval battles, and most of the islands and coasts charted by Pīrī were in enemy territory. Cephalonia switched hands twice – the Ottomans took it in 1479, but the Venetians recaptured it for Christendom in 1500. The map of the island of Rhodes would have been of military value to the besieging Ottoman forces (p. 137). Although the copy shown here was made many decades after the island was captured by the Ottomans, the first version of the *Book on Seafaring* was completed a year before the Ottomans finally took Rhodes from the Knights Hospitaller of the Order of St John. The main port city of the island is indicated in the north by a schematic set of buildings crammed together inside a yellow ring, and two towers guard the entrance to the port. The crosses around the harbour indicate hidden reefs, and the line of dots in the north-west a series of tiny islets.

اشکال نفس چجلیه بیان ایدر

سوزا چجلیه جزیره سی مالته برینه یازیلوب ومالته جزیره
اینوکجه صوده چجلیه برینه یازلمشلر

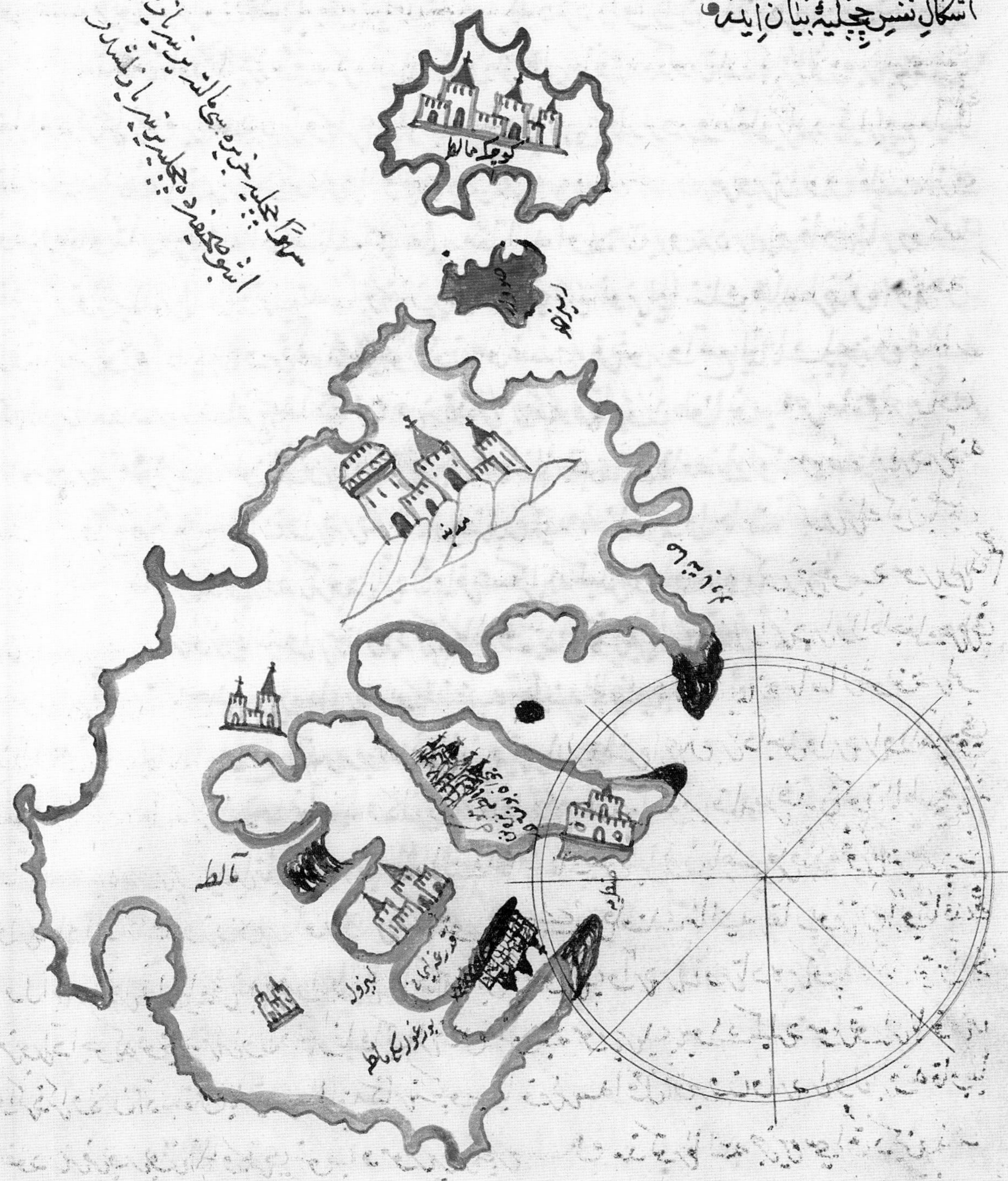

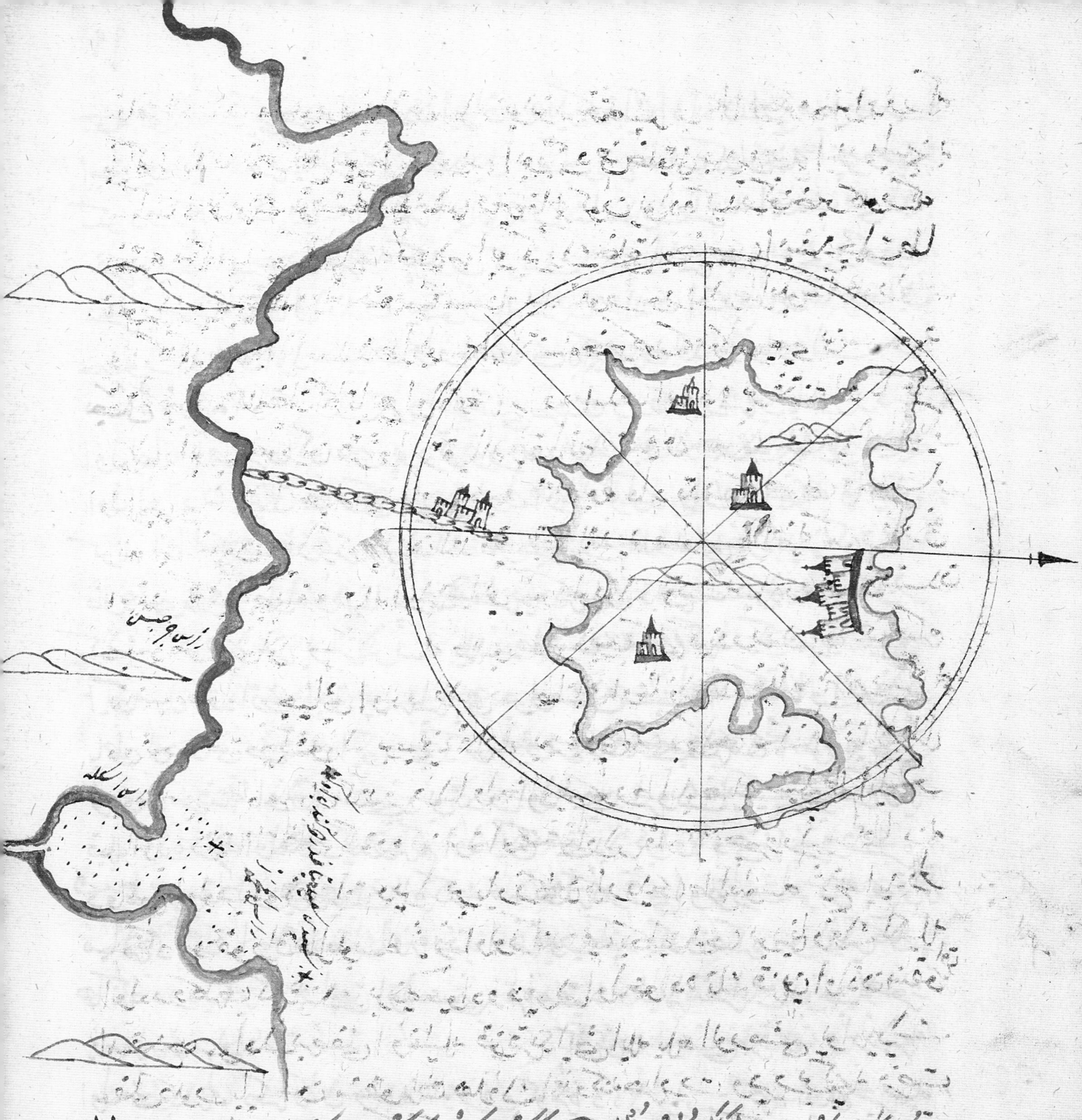

بو فصل طرابلس غربک کنارلرین بیان ایدر مغرب دیارنده طرا
بلوس قلعه‌سی یکی بنادر صحیح تاریخ بناسی کورد قلعه‌ده کوردم هر برج باروسی
سومه‌لرده قومیه یکرمی توتدی سلطانی مولانا عثمان زماننده مولانا عثمانک ابو بکر
ادلو براوغلی طرابلوسی بکیمش اول تاریخده طرابلوس علوفه قاسی بولدا اشدرمردن
اشد ور قورقه بولانا عثمان وفات ایدیجک مر بورک بیوک اوغلی مسعود اتا
سندن بر کون اول وفات ایدر مسعودک یحیا ادلو براوغلی قلور مولانا

The island of Jerba, showing a causeway from the Tunisian mainland, from the *Book on Seafaring* by Pīrī Reʾīs, copied 1587. Bodleian Library, University of Oxford, MS. d'Orville 543, fol. 98b.

The bays of the eastern shorelines of the island are much more pronounced than they are in reality.

When the Knights of St John were forced out of Rhodes in 1522, they chose Malta as their new base, and Pīrī's map of Malta is remarkable for its detailed depiction of the bays around the port of Valetta (p. 139). The teeth of the bay are precisely depicted, but again are disproportionate to the inland areas of the island, making them the focus of attention. The imposing churches reflect the Christian identity of the island, which would forever elude the efforts of the Ottoman sultans to capture it. The island of Küçük Malta (Turkish for 'Little Malta', present-day Gozo) and its castle are shown to the north, marked in bright red. In the text that prefaces this chart Pīrī mentions that Malta measures 65 miles in circumference, and has sixty villages and a castle situated atop a hillock some 4 miles inland. In the map, the castle dominates the inland areas of the island. He then surveys the navigation conditions in each of the major bays shown on the map, and concludes by describing the safe passages for ships between Malta and Gozo on both sides of the islet of Kamuna.

The *Book on Seafaring* is not only an Ottoman Turkish version of portolan charts, Italian *isolarii* and navigation guides. It is also a very personal text, where Pīrī narrates his life through the ports he has visited. This is partly because the charts were based on direct experience of the hazards of navigation, the identification of landmarks and correct compass bearings. As Pīrī wrote in the introduction to the second version:

> I have roamed the shores of the Mediterranean, Arabia and Europe, and through lands of Anatolia and the Maghreb
> And I have written, my friend, all that needs to be written about each and every thing.
> What sort of places they are, whether they are high ground or low,
> And I would have it known what place one will arrive at when disembarking and what the distance may be.
> Good reader, whenever I arrived at a port I would also make careful note of such things:
> Its anchorages, its wells, and all its landmarks, and thus all the details of the Mediterranean Sea.[9]

The text of the *Book on Seafaring* is replete with personal reminiscences, and much of what we know of Pīrī's biography is derived from the accounts of ports and

برنديز كنارلرينك وقلعه سنك شكلي بو رسمه در

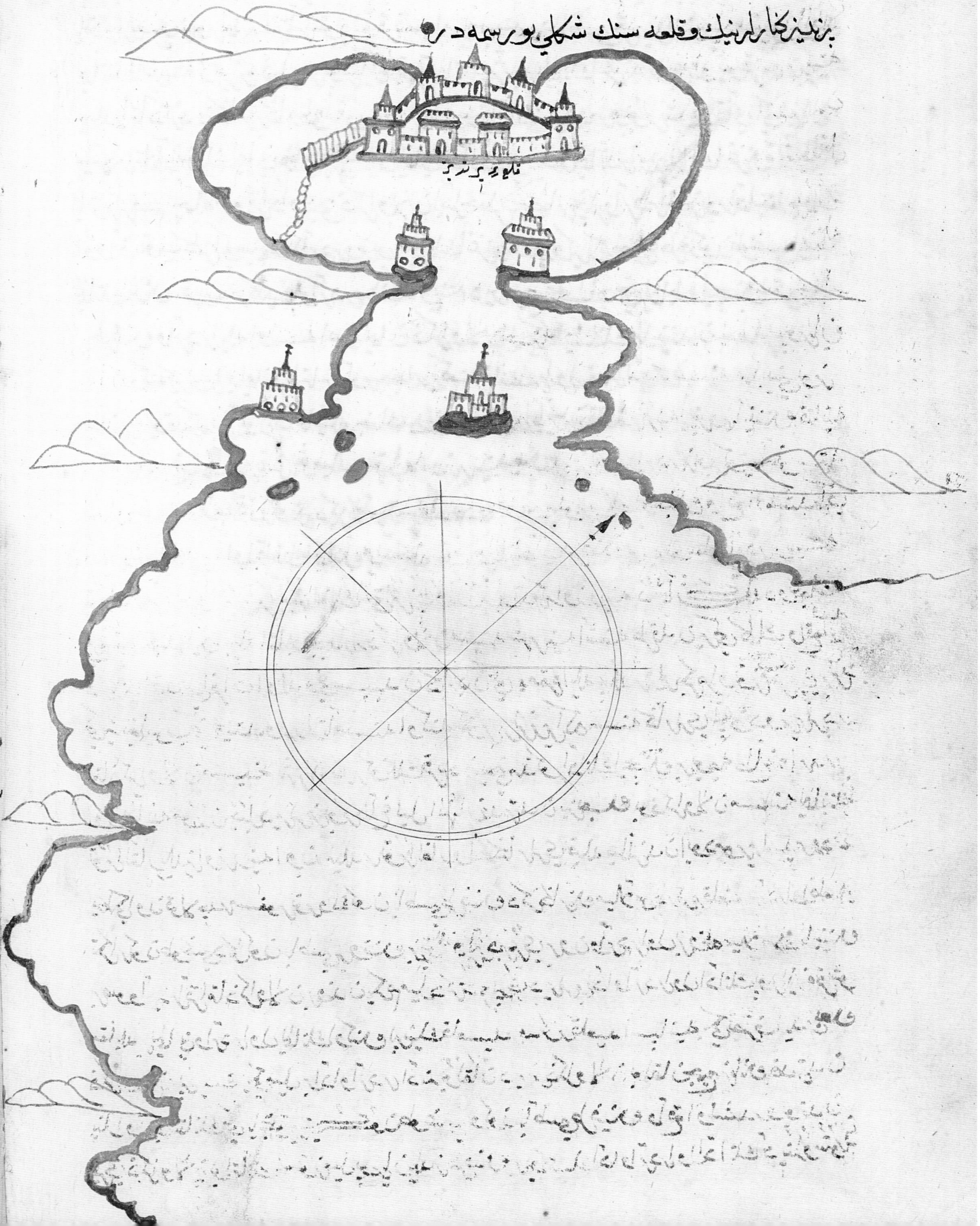

The port of Brindisi from the *Book on Seafaring* by Pīrī Reʾīs, copied 1587. Bodleian Library, University of Oxford, MS. d'Orville 543, fol. 79b.

islands, especially those along the North African coast. He raided Sardinia and Corsica with his uncle Kemal Reʾīs, whom he clearly idolized. He recalls how he and his uncle spent two winters in river anchorage near Bône in North Africa. The personal aspect is especially powerful in the account of the fertile island of Jerba near the coast of modern Tunisia, which was Kemal's pirate base (p. 140). The map shows a prominent causeway built by a sultan of Tunis, who wanted to subdue the island by connecting it to the mainland. The section closest to the island was then dug out by a local ruler, Shaykh Yaḥyā, in order to prevent mainland soldiers from marching over to the island. Pīrī dwells on the strange syncretic customs of the islanders. For example, when the members of the ruling clan drink water, they don't let the cup touch their lips; if it does, they break the cup. He also states that over the door of every house there is a cross to drive the devil off.[10] Above all, because the island is entirely surrounded by shallows that dry up twice a day, no ship unfamiliar with it can approach it. It is thus clear why, even with the best of charts, Jerba was a haven for pirates.

In this version of the *Book on Seafaring*, the treatment of architecture is generally schematic, as befits a volume with seamen in mind. A fortified tower stands for a fortress, a single gabled building for a town or a village, and columns falling or standing haphazardly represent ruined sites, mostly on the Syrian and Palestinian coasts. Nonetheless, the entrances to the ports can be shown in remarkable detail, as in the beautiful chart of Brindisi in southern Italy (opposite). As in other charts, the dark arrow on the wind rose indicates north, and the islets at the entrance to the port are distinguished by their bright colours. The chart marks the location of the series of towers guarding the entrance to the bay, and of the chain that blocks the entrance of the port itself, diagonally stretched from the city towards the southern shores of the bay. The walls, cathedral and castles of the city itself are merely schematic.

Initially, the *Book on Seafaring* was composed as a service for other captains of the Ottoman navy. Then, in 1524 or 1525 Pīrī happened to captain a ship that travelled from Istanbul to Rhodes carrying the Grand Vizier Ibrahim Pasha, who was sent to pacify the newly conquered province of Egypt. Pīrī's reliance on maps when plotting his course aroused Ibrahim's curiosity – there was perhaps something obsessive about Pīrī's relation to maps. The two started to converse about navigation and cartography for the rest of the journey, and Pīrī showed the Grand Vizier a complete copy of the *Book on Seafaring*, full of details on Mediterranean navigation but roughly sketched. Ibrahim advised Pīrī to compose

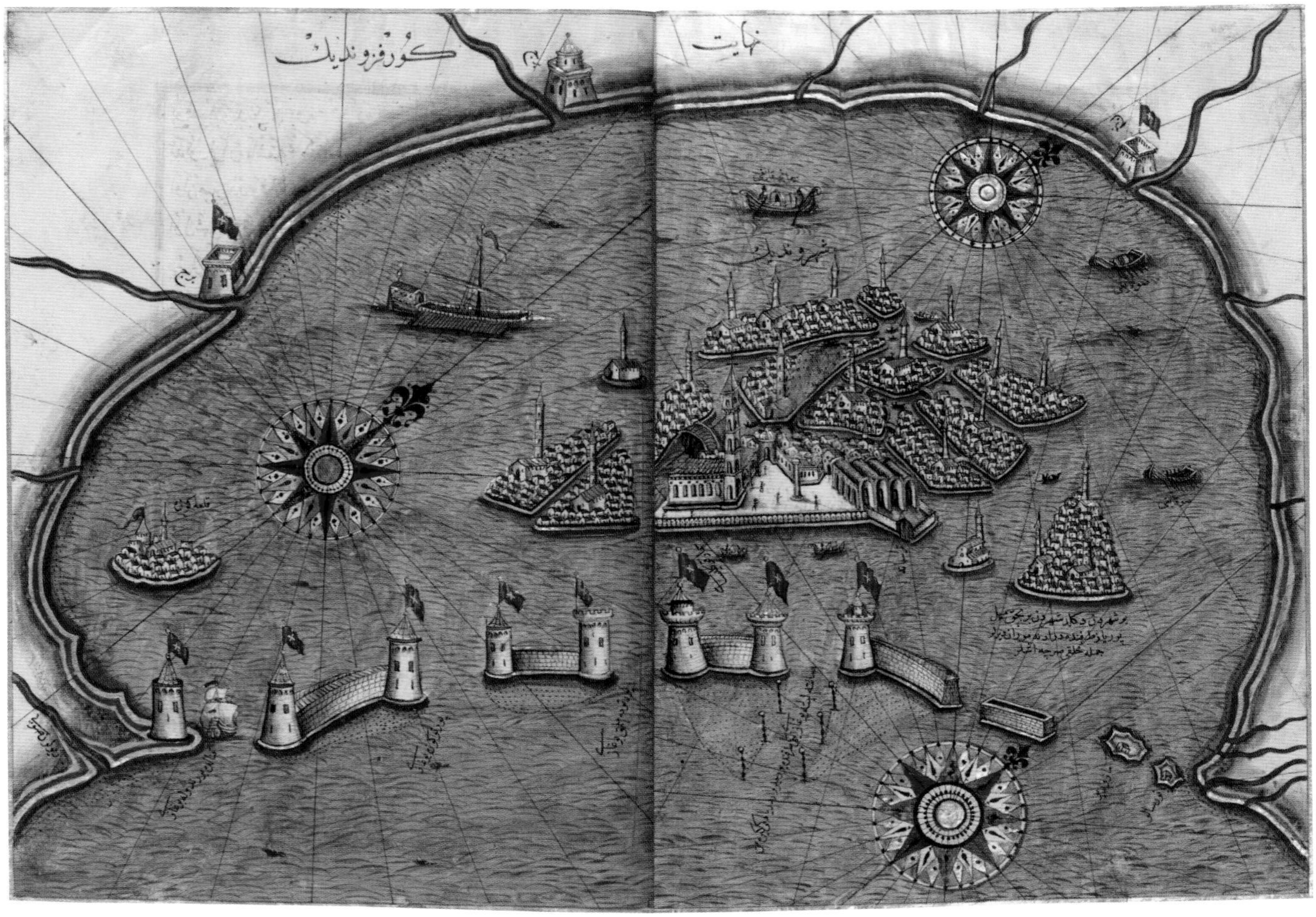

Map of Venice in the second version of the *Book on Seafaring* by Pīrī Reʾīs, copied *c.*1700. Walters Art Gallery, Baltimore, W.658, fols 185b, 186a.

a more polished version of the book, and present it to the sultan. That second version was finished by 1526, and presented to the young Sultan Suleiman shortly after his accession to the throne.

The second version turned out to be a lavish presentation copy. It begins with a long discourse, written in rhyming verses, on navigation, maps and European discoveries. The charts are adorned like miniature paintings, and incorporate elaborate town views influenced by European models. There are also new chapters and charts, with most additional material relating to Egypt, the Adriatic coast of Italy and the Gulf of Venice. The most striking new image is that of Venice, the Ottoman arch-enemy (above). Here again are vessels, flags and decorated wind roses, all reminiscent of the images on the world map. At the centre of the map is the bell tower of San Marco, which Pīrī notes to be the first landmark visible to pilots. Facing the bell tower across the public square is the naval arsenal, with its own fortified entrance. Other quarters of the city are separated from each other by canals, each quarter schematically represented as a host of buildings around a single church. This is undoubtedly Venice, but the illustration would not be a good guide to finding your way around town.

The Ottomans Face East

In the decades following Pīrī's *Book on Seafaring*, town views gained in popularity in the Ottoman court. Some of the most distinctive examples come from *Beyān al-Manāzil*, an illustrated campaign diary written by Maṭrāqci Naṣūh and presented to Suleiman in 1537.[11] Maṭrāqci was a jack of all trades, who served in the Ottoman court as an occasional miniaturist, Arabic translator, mathematician and sword-maker. Between 1533 and 1536 he was part of the retinue of Sultan Suleiman during his lengthy Mesopotamian campaign against the Shia Safavid dynasty. The *Beyān* is a diary of the campaign, illustrated with 130 miniature depictions of the major stops on the journey. The first is a majestic double-page bird's-eye view of the imperial capital Constantinople/Istanbul (overleaf). It is made in the tradition of bird's-eye views of Constantinople, first found in Buondelmonti's book of *isolarii*. The inlet of the Golden Horn cuts the image in two, with the Galata quarter and its distinctive tower on the left and the intramural areas of the main city on the right. Here we can easily identify some of the most familiar monuments of the city: Hagia Sophia, Mosque of Beyazid II, the newly constructed covered market (*bedestan*) and the Topkapı Palace. The Ottoman galleys sailing through the Golden Horn could have come from Pīrī Reʾīs's world map.

Maṭrāqci's originality comes to the fore when he depicts inland towns in Anatolia, Syria and Iraq, including Baghdad, Tabriz and Basra. Here town views are released from the naval context in which they originated, and are no longer limited to the port cities where the land meets with the sea. His famous town view of Aleppo has a distinctive Ottoman feel, and conveys to the viewer a sense of stability and peacefulness (p. 148). We have no earlier illustration of Aleppo from the Islamic world, and the act of drawing the city is itself a political act. Maṭrāqci's town view asserts the benefits conferred on the city since it came under Ottoman rule a mere two decades earlier. The city is recognizable as Aleppo by the mighty citadel, built in the twelfth century. The rows of structures are generic, with the minarets indicating Muslim religious buildings. Strikingly,

Town view of Constantinople/ Istanbul from the *Compendium of Stages* (*Beyān al-Manāzil*), an illustrated campaign diary completed by Maṭrāqci Naṣūh in 1537. Topkapı Palace Museum, Istanbul, Turkey / Dost Yayinlari / Bridgeman Images, T. 5964 / XYL161187.

Detail from a town view of Aleppo, from the *Compendium of Stages* (*Beyān al-Manāzil*), an illustrated campaign diary completed by Maṭrāqci Naṣūh in 1537. Private collection/ Bridgeman Images, T. 5964/ CHT779776.

the style of the minarets is distinctly Ottoman, although in 1537, when this map was made, no mosque had yet been built in this style. Moreover, there are no indications of churches and synagogues, even though sixteenth-century Aleppo had many of these. As rightly observed by Heghnar Watenpaugh, Maṭrāqci represents Aleppo 'not as it was, but as it ought to be'.[12]

Increasingly, the Ottomans were facing east. Fighting in Mesopotamia was followed by concerted attempts to rival the Portuguese penetration of the Indian Ocean. By the 1530s the Ottoman navy was building galleys in Suez, and had also established a naval base in Basra. Aden, ever the gateway to Indian Ocean trade, was captured by the Ottomans in 1538. But fighting in the Indian Ocean required different skills and different vessels than in the Mediterranean. It also required new geographical knowledge of the kind offered by Pīrī. From the point of view of the Ottoman court, Pīrī's 1513 world map was useful for its depiction of the Indian Ocean, while the New World was still a curiosity. This explains why the map was torn up; the surviving portion of the map, covering the Atlantic, was of no military value and put in the libraries. The sections of the Indian Ocean that Pīrī derived from Portuguese maps were put to practical use.[13]

In 1547 Pīrī Re'īs, now commander of the Alexandria squadron and the most senior naval officer in Egypt, was called to face off the Portuguese in the Indian Ocean. In the versified introduction to the second version of the *Book on Seafaring*, written twenty years earlier, Pīrī acknowledged the challenges of navigating in the Indian Ocean. In the closed Mediterranean, distances are short and an experienced pilot can always correct a faulty compass or an imprecise chart. But sailing in the larger Indian Ocean was more dangerous, and the Portuguese had mastered it in a way that elicited Pīrī's admiration. Nonetheless, Pīrī's first foray into the Indian Ocean was a success, and in February 1548 he managed to dislodge a Portuguese force that had taken the key port of Aden in the previous year. But his campaign against Hormuz five years later failed, and he paid for that failure with his life. Was Pīrī's professional respect for the Portuguese seamen the cause of his downfall? Should Pīrī, now in his seventies or even eighties, have been braver and persevered in the siege on Hormuz in 1553 and in taking on the Portuguese reinforcements arriving from India? Perhaps. But we should remember that his *Book on Seafaring* covered only the Mediterranean. In the treacherous shallows of the Persian Gulf, he was a fish in the wrong waters.

The sixteenth century was the age of Ottoman expansion, and by 1600 it was the largest empire in the Old World, stretching from North Africa to Iraq and from

زیکسو رصد خانهٔ مختصر
نمودند نزدیکی آن مقر
درو پانزده اهل علم گزین
شدند ازپی خدمت تقی دین
پس انکه ترصید مریک ازان
بشد شیخ ما فاضل نکته دان

Astronomers at the observatory in Galata, from Sayyid Luqmān's official history of the Ottoman court, 1581. Istanbul University Library, Istanbul, Turkey/Bridgeman Images, F 1404/XYL155434, fol. 57a.

the Balkans to Egypt. Cartographers provided practical information in military contexts, charting siege maps and the precise fortifications and entry points of enemy port cities. Ottoman cartographers like Pīrī also provided a vision of the world that shaped imperial policy.[14] This was a vision that had only loose ties with the medieval Islamic tradition. Al-Istakhrī's maps were being copied but more out of nostalgia; the mathematical geography of al-Khwārazmī and al-Idrīsī was of little interest. Instead, Ottoman map production was a synthesis of the empire's interaction with map genres that originated in Europe, including portolan charts, *isolarii* books, reconstructions of Ptolemy's *Geographia* by Greek scholars and Italian town views. Pīrī's great achievement was to take these different genres and make them his own, naturally Ottoman and Turkish but interwoven with his own observations as a practising seaman. Through his charts, the seas could be imagined as Ottoman seas.

Pīrī was not the official cartographer of the Ottoman empire. Unlike the Spanish, the Ottomans did not establish an official body where geographical knowledge was produced, regulated and censored. But in 1579 the Ottoman sultans did construct a short-lived observatory in the Galata Tower. The scene within the observatory is depicted in a miniature illustration accompanying an official history of the Ottoman court, a work presented to Sultan Murad III in 1581 (opposite). In the background, some dozen astronomers and their tools of observation can be seen (of which more will be said in Chapter 6). At the front there is a fine globe, on which the outline of the continents is easily discernible. The Atlantic coast of South America, as in Pīrī's world map, and an oversized Africa can be seen. The Ottoman domains are, of course, at the centre. The notion of representing the Earth as a globe, which is so familiar to us now, was new in Islamic cartography. A globe had been necessary only once the New World was discovered. Previously, since the entire inhabited world was limited to one hemisphere, there was no need for a three-dimensional representation of the Earth. Unfortunately, this sixteenth-century Ottoman globe did not survive, nor has any other globe of this type. But the message of the image is clear: the observatory scholars, dressed in robes and turbans and wielding their quadrants and astrolabes, are taking part in an age of discovery.

صفة دائرة يستقبل الى الكعبة المشرفة
ب وصفة ٥ يستقبل القبلة في الصلوات

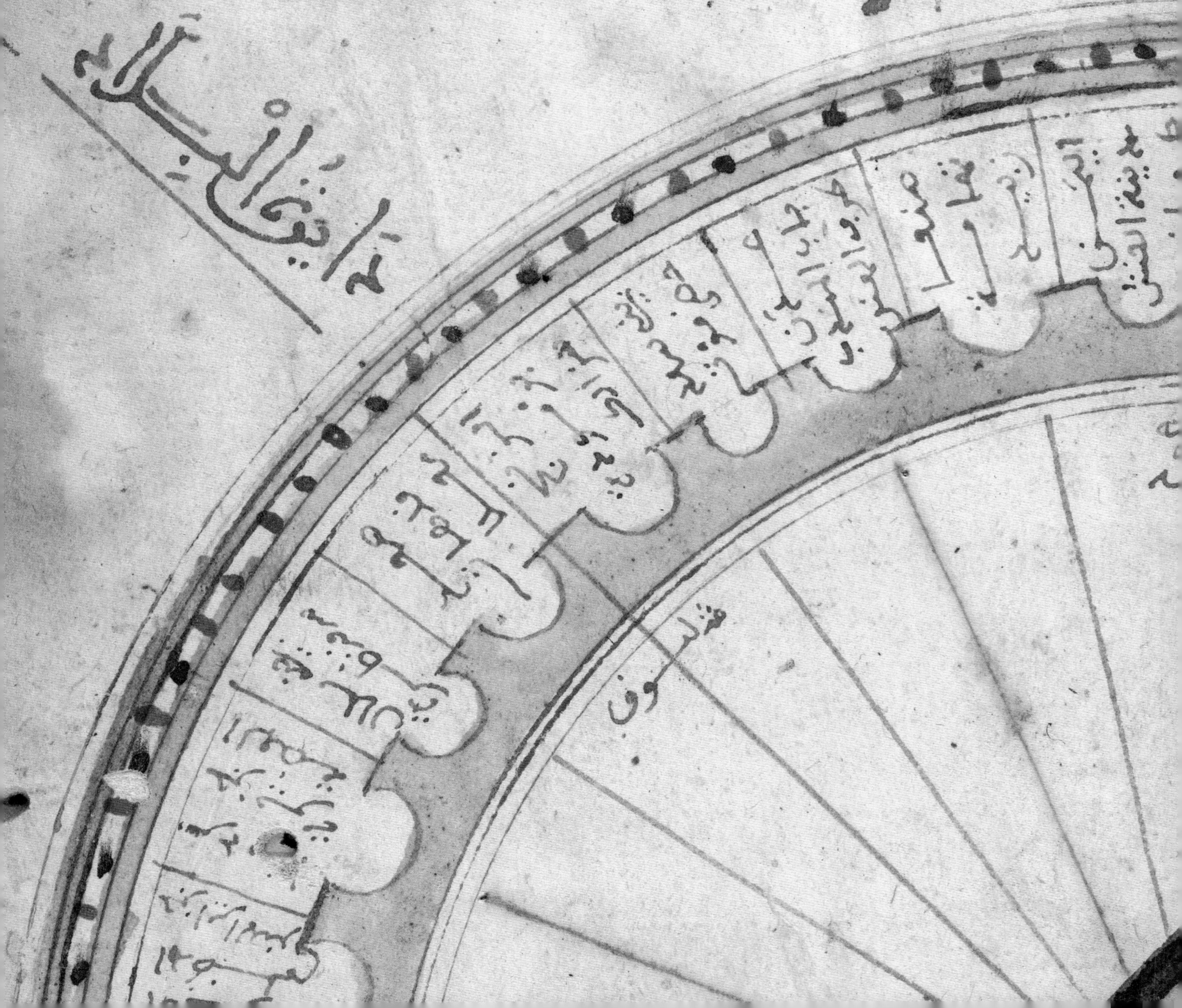

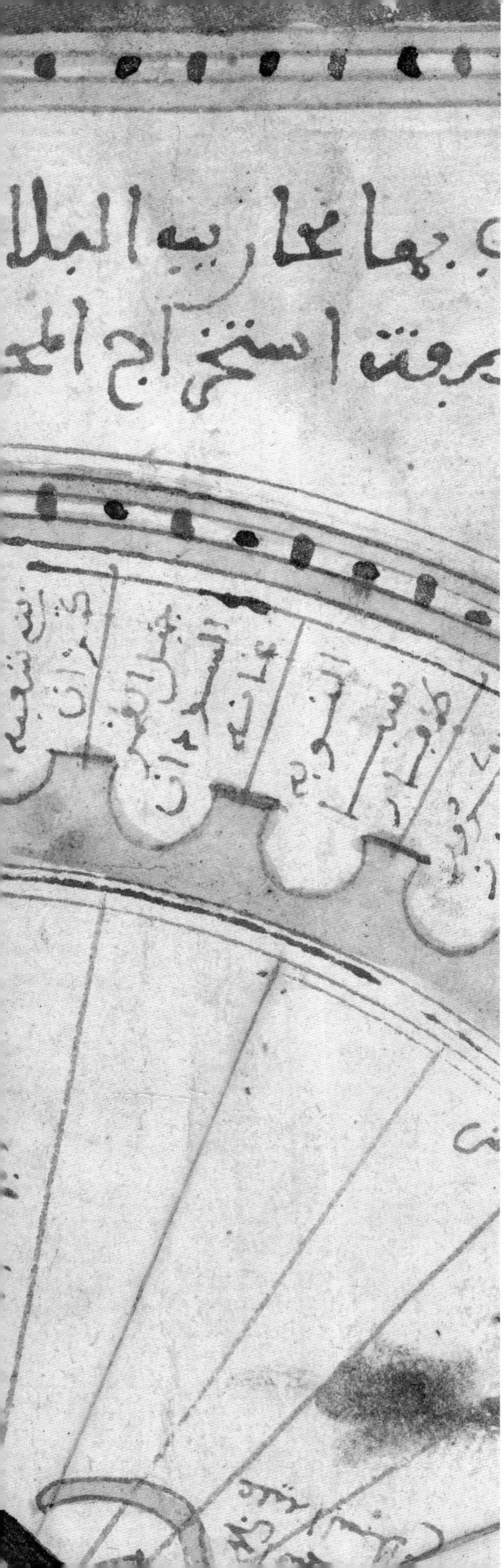

CHAPTER 6

An Astrolabe for a Shah, or Finding the Direction of Mecca in Safavid Isfahan

The face of an astrolabe made for Shah ʿAbbās II of Isfahan, by Muḥammad Muqīm Yazdī, in 1647/8. © History of Science Museum, University of Oxford, 45747.

IN THE YEAR 1057 of the Islamic era, corresponding to 1647–8 CE, the court astronomer of Shah ʿAbbās II, the Safavid ruler of Iran, presented the young emperor with a magnificent instrument, as mathematically accurate as it was exquisitely executed. The face of the instrument was a finely finished astrolabe, beautifully inscribed with the name of the sultan on it (opposite). On the back of the instrument was a rotating sighting device and the signatures of the astronomer and the craftsmen who had produced it. Also on the back, in the top right quarter, was a quadrant engraved with lines allowing the user to find the direction of Mecca by the altitude of the sun at their location (overleaf). At the base of the instrument, below the removable plates of the astrolabe, were the latitude and longitude coordinates of forty-six different cities in the Safavid domains, as well as the numerical value, in degrees, of the direction of prayer towards Mecca in each of these cities (p. 158). The instrument exuded a sense of control over time and space, integrating celestial maps with the sacred geography of Islam and the political authority of the Safavid empire. It was truly an astrolabe for a shah.

Astrolabes, the most sophisticated time-telling devices of pre-modern societies, were also used for calculating the rising and setting of the sun, as well as for casting horoscopes, and had been produced in Islamic contexts since the ninth century.[1] The top plate of the astrolabe is a star map, where the tips of the brass pointers indicate a select number of prominent stars. The inner circle at the upper half of the star map traces the movement of the sun across the sky in the different months of the year. This star map can be rotated, and the rotation represents the daily rotation of the skies relative to an observer at a particular geographic latitude. If the shah wanted to tell the time, whether by night or by day, all he had to do was to observe the altitude in the sky of either the sun or any prominent star. He would then rotate the top plate so that the location of the star he observed aligned with the lines of altitude inscribed on the plate beneath. The lower plate shown here is specific to the altitude of the shah's location – his imperial capital, Isfahan. He would then be able to read the time, remarkably precisely, from the numerals engraved on the outer rim. If he happened to travel elsewhere in his realm, he could replace the bottom plate with another plate that reflected the altitude values in another latitude.

While astrolabes had been produced in the Islamic world since the Abbasid period, the quadrant on the back of the instrument was an ingenious Safavid invention (overleaf), which allowed the user to find the direction of Mecca by

ظل اقدام مستوی
ظل اصابع مستوی

A *qibla* quadrant for finding the direction of Mecca, on the back of the Shah ʿAbbās II astrolabe, top right corner, made in 1647/8. © History of Science Museum, University of Oxford, 45747.

observation of the sun's altitude. In the instrument made for ʿAbbās II, the names of fourteen cities are written on the outer edge of the quadrant, and a curved line leads from the city name towards the centre of the instrument. Each of the lines shows the altitude of the sun in that location when the sun is in the same direction as Mecca. Since the sun's position in the sky changes throughout the year, the user would also need to know the month or, in this case, the zodiacal sign during which the observation was being made. The movement of the sun is represented by the concentric quarter circles, each aligned with a label for one of the twelve signs of the zodiac. The names of the twelve zodiacal signs are written on the lower part of the quadrant.[2]

Once it has been explained, it is quite straightforward to use this quadrant to find the direction of Mecca. First, you have to find the curved line – the one representing the altitude of the sun when it is in the direction of Mecca – that corresponds to your location. If you were Shah ʿAbbās II, that would be your imperial capital of Isfahan, indicated by the eighth line counting from the top. You then find the concentric quarter circle that represents the time of the year, based not on Islamic months but on the signs of the zodiac. The intersection of the curved line for Isfahan with the circle showing the time of year gives you the angle in degrees, between 0 and 90, of the sun's altitude when it is in the same direction as Mecca. For example, in Isfahan at the beginning of Aquarius (i.e., for a short period from 21 January), the lines intersect around 25 degrees of altitude. Only then do you need to actually look at the sky and measure the sun's altitude with a sighting device, called an alidade, which is attached to the back of the astrolabe. In Isfahan in late January, for example, you would wait until the sun is at 25 degrees, when you know that it is in the same direction as Mecca, and therefore that is the direction you need to face for your prayers.

The circular table in the recessed area holding the plates (overleaf) is another Safavid innovation, and one that highlights a fascination with mathematical methods of determining the direction of Mecca. This circular table has an outer and an inner ring, each with four rows. The top row in each of the rings represents the name of the city, the second row the longitude of that city, the third row its latitude and the fourth the direction towards Mecca from that city. The Safavid instrument took into account not only the known coordinate values of both Mecca and Isfahan, but also the curvature of the Earth. For the Safavid capital of Isfahan, unsurprisingly represented at the top of the outer ring, the values are 86° 40' of longitude (measured from a prime meridian in the Canary

Radial base of the Shah ʿAbbās II astrolabe, with coordinates and *qibla* values for forty-six cities, made in 1647/8.

Islands), and 32° 25' of latitude north of the equator. The direction of prayer towards Mecca is given at 40° 28' west of south, which is equivalent to 220° 28'. This is a pretty precise computation of the direction of prayer from Isfahan: currently websites that use the latest technology give values from 223° 94' to 226° 06', depending on the method of calculation.[3]

In truth, Shah ʿAbbās didn't need this instrument to tell him either the time of day or the direction to Mecca, and not only because he was the shah and could always call on a servant. At the time, Isfahan already had precise sundials and water clocks, as well as several mechanical clocks installed by European artisans. Moreover, the direction of prayer was well known to all residents of Isfahan, and particularly to court officials and religious scholars. By the tenth century Muslim scientists had already calculated the direction of Mecca from a multitude of localities, including Isfahan, and in the fifteenth century a team working at an observatory in Samarkand perfected these calculations. Although the instrument made for the shah was as accurate as was possible for the seventeenth century, it did not present anything that was not already known.

So how should we read this magnificent masterpiece of metalwork, art, science and piety? First, the instrument and its inscriptions made a statement about power. The dedication on top of the astrolabe calls on God to perpetuate the empire of the shah, and to 'cause his justice and his benefits to spread over the worlds while the spheres revolve and the planets continue in their courses'.[4] On closer observation, his name, 'Sultan Shah ʿAbbās the Second', in Arabic, can be seen interwoven in the rotating map of the sky, just inside the circle showing the movement of the sun. The power of the emperor is visually and textually projected as an integral part of the celestial order, his reign seen to be as natural as the celestial orbits.

On another, perhaps deeper, level, this instrument's representation of the direction of Mecca should be read as an ideological statement about the correct interpretation of God's will. In Safavid Iran, which espoused the Shia school of Islam, the makers of this instrument were part of the battle between competing camps of Shia scholars, representing two competing theological outlooks. While Shah ʿAbbās may not have had any practical use for this instrument, it signalled his support for a strand of Islamic thought that valued philosophical, astronomical and mathematical enquiry, and his opposition to traditionalist scholarship which answered all religious questions by referring to scripture and layman knowledge of the world. The message here is that the direction of

Mecca could be known through astronomical observations and mathematical calculations, and that acquiring this astronomical and mathematical knowledge is an act of piety. What mattered was not only mere knowledge of the direction of Mecca but the ability to find the sacred direction on one's own, through one's own powers of reasoning and observation. This instrument enabled its user to actively participate in the quest for determining God's will rather than slavishly imitating the opinions and judgements of others.

The *Qibla* towards Mecca

This final chapter traces the way in which Muslims have represented their sacred geography over the centuries, culminating with the unprecedented flourishing of instruments for finding the direction to Mecca in early modern Iran. Unlike the previous chapters, we shall not follow the makers of the astrolabe made for Shah ʿAbbās II, even though their names are known to us.[5] Instead, our interest here is in the way in which visual representations of the direction of Mecca reflected and projected different religious attitudes within the great plurality of the Islamic world. Specifically, maps for finding the direction of Mecca represented opposing ideas about the value of scientific concepts and instruments: some Muslims harnessed and incorporated the precision of scientific technology in the service of their religious identities, while others were less concerned with precision and more with lay accessibility to knowledge and with attachment to past traditions.[6]

Finding the direction of Mecca is a ritual obligation on all Muslims. The Qur'an specifies that Muslims should pray in the direction of the Kaaba in Mecca (Qur'an 2: 144–50). In the Qur'an, this prayer direction appears to distinguish Muslims from other religious communities, undoubtedly referring to Jews, who pray towards Jerusalem, and Christians, who pray towards the east. In Arabic, the direction of the Kaaba is called *qibla*, a term which has subsequently been used in all languages of the Muslim world, and even by Arabic-speaking Jews and Christians to refer to their own sacred directions of prayer. Post-Qur'anic tradition then amplified the importance of the sacred direction beyond prayer, and required various ritualistic activities to be performed in the direction of the Kaaba, including burial of the dead, recitation of the Qur'an and ritual slaughter; in contrast, bodily functions are to be performed in a perpendicular direction to Mecca. The *qibla* was therefore inscribed in the physical landscape of Muslim societies, in cemeteries and above all in mosques, where the prayer niche (*mihrab*) was to be oriented in the direction of the Kaaba.

While praying towards the Kaaba is a ritual obligation, finding the direction of the Kaaba is an act of religious interpretation. The verses directing Muslims to pray in the direction of Mecca were revealed to Muhammad while he was in exile in the town of Medina, north of Mecca; he therefore prayed due south. In the generations after Muhammad's death, many early Muslims simply emulated his actions and prayed due south wherever they were. Others deduced the direction of Mecca from the direction of the road that led towards Arabia in any given locality. Most commonly, however, Muslims used what has been dubbed folk astronomy. This meant that Muslims observed astronomical phenomena on the horizon, such as the risings and settings of celestial bodies, which they then aligned with the direction of Mecca. The winter solstice, the shortest day of the year, held special importance: Muslims in Iraq often assumed the direction of prayer to be the direction of the sunset on that day; in early Islamic Egypt the direction of Mecca was aligned with the direction of the winter sunrise.

So, when Muslims started building mosques all over the Middle East in the eighth and ninth centuries, the orientation of the prayer niches was as diverse as the localities that produced them. The directions that Muslims followed were always approximate, and at the time architects did not have recourse to mathematical calculations. This meant a plurality of orientations even within the same city.[7] The Great Mosque at Córdoba, built in 784, has its prayer wall in the direction of the south-east, at 152 degrees, even though Andalusi scientists were well aware that Mecca was almost due east, at 100 degrees. It is likely that the mosque was laid out in accordance with the existing Roman-era street plan, thus adapting itself to local conditions. In eleventh-century Samarkand, in Central Asia, mosques associated with the Hanafi legal school were oriented due west, following the road leading to the Arabian Peninsula, while mosques associated with the Shafiʿi school were oriented due south, following the example of the Prophet in Medina. Some mosques were oriented to the south-west as a compromise between the two schools, while the Friday mosque of the city was oriented towards the setting of the sun at the winter solstice.

The first cartographic representations of the direction of the *qibla* were similarly approximate, and emphasized observation of the rising and setting of stars as the primary means of finding the *qibla*. Such a schematic diagram of the direction of the *qibla* is found in a late medieval copy of a geographical treatise by al-Muqaddasī, a follower of al-Iṣṭakhrī working in the tenth century. Even if this illustration was not original to al-Muqaddasī's treatise, it appears to reflect

السادس لان آخر الذي هو اول هذا واخر الذي يلي الجنوب حيث وقع الطرف الاقصى الشمالي
من الاقليم الذي يليه وهو السادس وذلك سمت خوارزم وطرازبند شرقا وغربا ووقع طرفه
الاقصى الذي يلي الشمال اقاصي ارض الصقالبة واطراف الترك الذي يلي خوارزم من جهة الشمال ووقع
ووقع وسطه في بلاد اللان ... بلا مدن معروفة

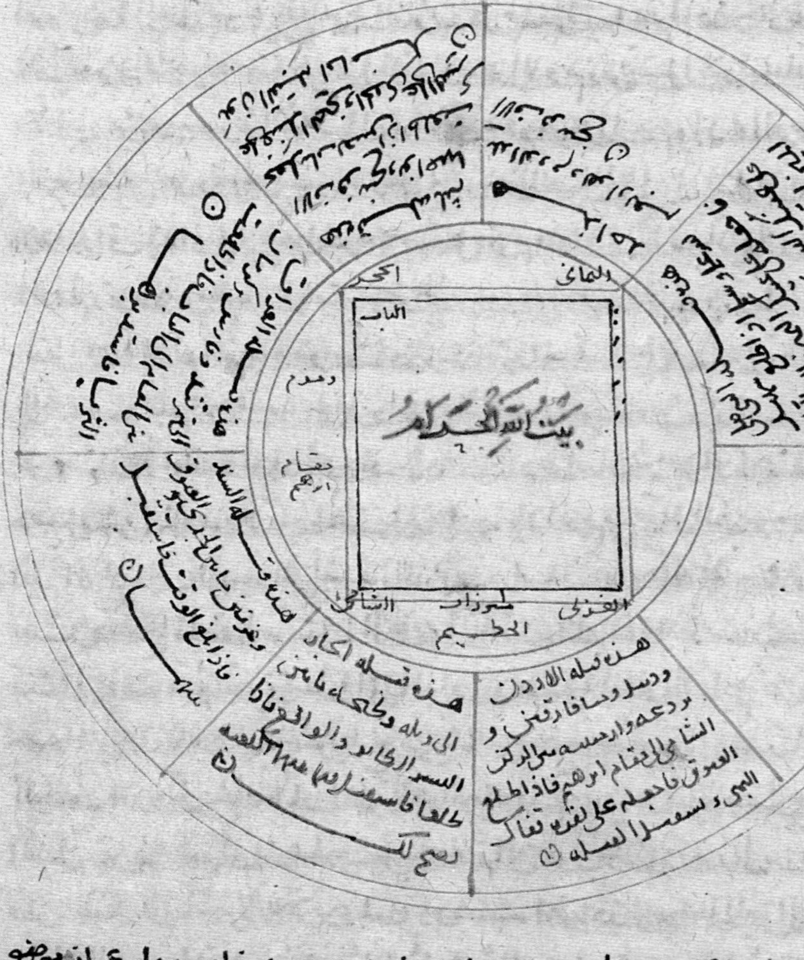

وقال عبد الله بن عمرو الدنيا مسيرة خمسمائة سنة اربعمائة سنة خراب ومائة عمران مواضع
المسلمين منها سنة ۞ وعن ابي الجلد قال الارض اربعة وعشرون الف فرسخ السودان
اثنا عشر الف فرسخ والروم ثمانية الاف فرسخ وفارس ثلثة الاف والعرب الف فرسخ ۞
ذكر مملكة الاسلام اعلم ان مملكة الاسلام حرسها الله تعالى ليست مستوية

مطلب جليل مهم
معرفة الدنيا والارض بالفراسخ

Qibla diagram from al-Muqaddasī's *The Best Divisions for Knowledge of the Regions*, copied 1494. © bpk Bildagentur / Staatsbibliothek Berlin, Orientabteilung, MS. Sprenger 5 (Ar. 6034), fol. 34.

an early Islamic scheme of sacred geography (opposite).[8] The Kaaba is shown in the middle of the diagram as a large square, labelled 'Sacred House of God'. The south-eastern wall of the Kaaba is shown at the top, and the Black Stone, the focus of devotion, in the eastern corner, on the top left of the structure. In the Qur'an and in Islamic tradition, Abraham and Ishmael are said to have built the Meccan temple at God's command, and Maqām Ibrāhīm, where Abraham and Ishmael stood when building the Kaaba, is indicated to the left of the building. As the image demonstrates, prayer was to be directed towards the Kaaba itself, and allows the believer to imagine themselves facing a particular wall, as if within the precincts of the Holy Mosque surrounding the Kaaba.

The circle surrounding the Kaaba is divided into eight sectors, each indicating the direction of prayer from one of the regions of the tenth-century Islamic world. All the regions in the same sector share roughly the same *qibla*. For example, the bottom-right sector, just below the northern corner of the Kaaba, indicates the direction of prayer for regions and cities located due north of Mecca, from al-Urdun (present-day Transjordan) to Armenia. The text of the diagram explains that the direction of prayer in these regions has been determined by observing the rising of Capella, one of the brightest stars in the sky. Next to it, the bottom-left sector represents the direction of prayer in ports along the western coasts of the Arabian Peninsula. In these areas, the direction of prayer has been determined by observation of the rising of two stars known as Flying Eagle and Alighting Eagle.

While this scheme provides only an approximate division of the world into eight sectors, Abbasid-era mathematicians were already calculating precise *qibla* directions, measured in degrees and minutes. The first mathematical determination of the *qibla* known to us comes from Baghdad in the early ninth century. In the tenth century, the geometer al-Buzjānī approached the challenge of finding the direction of Mecca as a problem in spherical trigonometry. He computed the value of the angle towards Mecca by calculating the angles of a spherical triangle whose points are Mecca, the North Pole, and the desired locality. In order to find the direction of prayer, one needed to know the longitude and latitude coordinates of both Mecca and one's own location, as well as one's distance from Mecca. Then one needed a table of sines, tangents and cotangents and to apply the law of sines to spherical trigonometry.[9] This was not a trivial mathematical problem, but in the centuries that followed the most advanced scientists of the age produced tables of *qibla* values for hundreds of cities all over the Muslim world.

The new mathematical knowledge, however, led to an unexpected problem: the *qibla* values calculated by the scientists did not match the physical orientation of the mosques founded by earlier generations of Muslims. In Egypt, for example, the early Islamic mosques were oriented towards the sunrise at the winter solstice, an orientation that differed by at least 10 degrees from the calculations based on coordinates of longitude and latitude. In the delta region of northern Egypt, where the calculated orientation was south-east, most of the prayer niches faced due south, as if following the orientation of Syrian mosques. Al-Dimyāṭī, a twelfth-century scholar from the delta, attempted to resolve this discrepancy by surveying the different methods of finding the *qibla* known to him. While he acknowledged the validity of the division of the world into sectors and the observation of the rising and setting of stars associated with folk astronomy, he argued that the *qibla* should ultimately be found through mathematical means and that there could be only one correct *qibla* in any region, rejecting the pluralistic multiplicity of the early generations.[10]

Al-Dimyāṭī illustrates his argument through a diagram (opposite), which has the rectangular Kaaba at the top left, oriented to the south, with the Black Stone shown in the eastern corner of the structure. The precise (*samt*, or azimuth) direction of prayer from Cairo is shown by the top line on the right. Below the azimuth line of Cairo are lines indicating the direction of prayer from Jerusalem, Damascus and Aleppo, in clockwise order. The squiggly line shows the pilgrimage route from Egypt and Palestine, and allows the user of this diagram to adjust their prayer direction as they proceed towards the Kaaba. Al-Dimyāṭī gently brushes aside the rough division into segments adopted by earlier generations, and ascribes the incorrect orientation of early Islamic mosques to the incomplete Islamization of Egypt at that time. Proper Islamic practice, he seems to suggest, is to find the *qibla* of one's location through all means possible and to disregard the precedent set by earlier generations.

The Late Middle Ages saw new technologies that were capable of determining time and space more precisely. By the thirteenth century, many mosques in Egypt and Syria employed a professional timekeeper, called *muwaqqit*, who was responsible for astronomical observations with the purpose of determining the times and direction of prayer. Mosques were also installed with complex sundials and water clocks. In Iran, the Shia polymath Nasir al-Dīn al-Ṭūsī (1201-1274) supervised the building of a large-domed astronomical observatory funded by the new Mongol rulers in their capital Maragha, and produced new sets of data on the

Illustration of the *qibla* directions from Egypt and Syria. From al-Dimyāṭī's treatise on the sacred direction, copied 592/1196. Bodleian Library, University of Oxford, MS. Marsh 592, fol. 88b.

وهذه صورة ذلك والعجب العجاب
من رضي المخالفة سبيلهم فما
استوى والله تعالى يلهم
الرشد من يشاء من عباده
بفضله

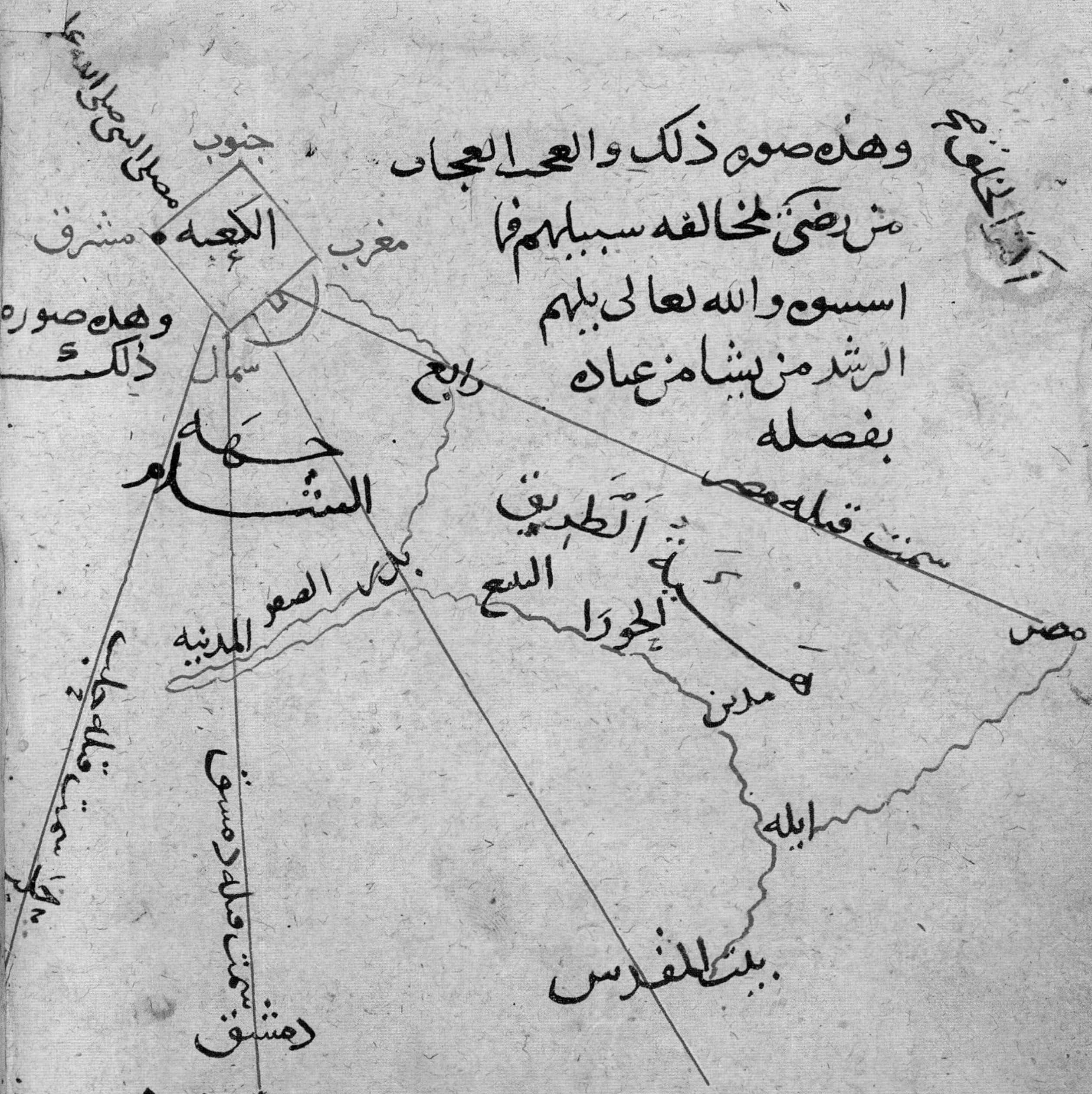

الباب التاسع في الادلة
الشرعية التي نصبها الله تعالى لعباده يستدلون بها
على سداد استقبالهم واستقامة طرقهم في اسفارهم

movements of celestial bodies. His measurements were improved in the fifteenth century by a group of astronomers working at a new observatory in Samarkand, an observatory that still stands today.

Perhaps the most important new piece of late medieval technology was the magnetic compass, whose revolutionary implications for navigation were discussed in the previous chapter. The application of compass bearings had dramatically altered the ability to determine the *qibla* precisely from any location. The cartographic result of that shift could be visually spectacular, as shown in a beautiful *qibla* diagram found in the portolan atlas made by the Sfaksī family in the sixteenth century, held by the Bodleian Library (opposite).[11] The familiar structure of the Kaaba is shown in the centre of the diagram, oriented to the south, with the Black Stone near the eastern corner. Maqām Ibrāhīm is shown in the bottom left alongside the Well of Zamzam, both facing the north-eastern wall of the Kaaba. The structures are far more ornate than in the simple rendering found in al-Muqaddasī's treatise. As in a map for navigation, the Kaaba is the centre for rhumb lines radiating into a 32-point wind rose. Each of the thirty-two points of the wind rose leads to a visual prayer niche, a *mihrab*, and in each of these prayer niches can be found the names of three regions or cities. For example, Baghdad, Kufa and Basra are grouped together at 8 o'clock while Fes and Marrakech are grouped together at 3 o'clock, due west.

As a result of the introduction of the compass, the map-maker was able to subdivide the limited number of sectors found in earlier diagrams into ever-thinner slices. The compass also obviated the need to search for the rising or setting of specific star groups in order to find one's orientation. The incorporation of this diagram into a navigation atlas suggests that it may have been used by travelling sailors wishing to adjust their prayer direction while coasting along the shores of the Mediterranean. But it is also noticeable that the localities are uniformly spread around the Kaaba and that there are no numerical values attached to these. Moreover, another copy of the same atlas has the distribution of localities along the ring in a substantially different order, suggesting that the *qibla* directions here are only meant as rough approximations. Overall, the effect is primarily decorative and ritualistic and, despite the compass bearings, the diagram does not aim for mathematical precision.

Qibla chart from a portolan produced in a family workshop in Sfax, in present-day Tunisia, 1571–2. Bodleian Library, University of Oxford, MS. Marsh 294, fol. 4b.

The Safavids and the *Qibla*

The *qibla* quadrant made for Shah ʿAbbās II, which reflected the flourishing production of astrolabes and *qibla* indicators under the Safavids, had its roots

ثم ايرة يعرف بها محاريب البلاد وصفة كل استقبال الى الكعبة المشرفة
وقد تقدم معرفة استخراج المحاريب وصفة كل استقبال اليمانية الطواف.

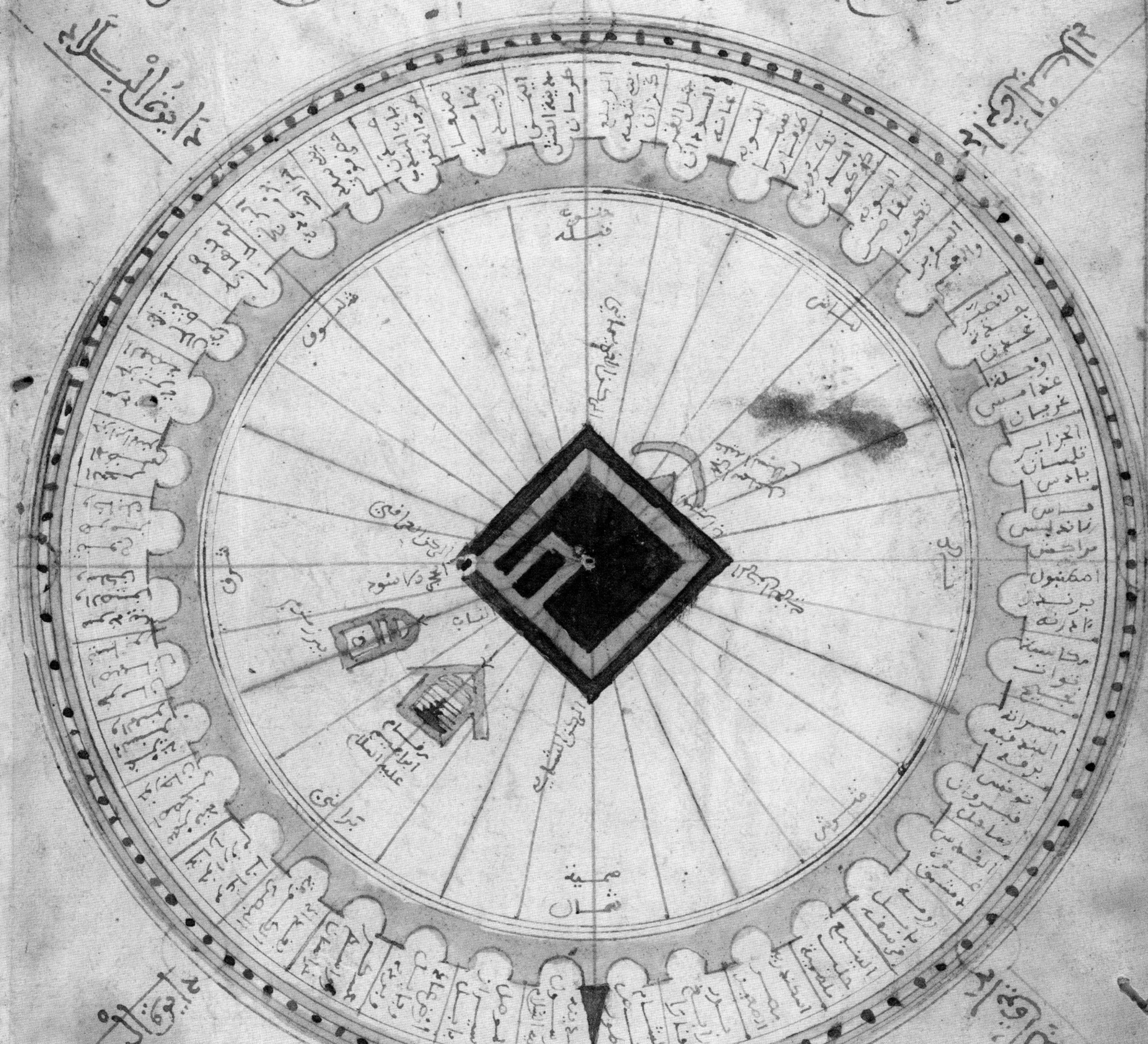

قال الله العظيم في محكم كتابه الحكيم وحيث ما كنتم فولوا وجوهكم شطره.

in the ideological turmoil that followed the establishment of the dynasty in 1501. The founder of the dynasty, Shah Ismail, was a charismatic leader of a mystical Sufi group in northern Iran, in what is today Azerbaijan. Claiming for himself divine descent and drawing on the heroic images found in the rich Persian epics, Ismail managed to defeat the many Timurid princes who were in control of the various cities of Iran. Within a decade, he forged a Persian-speaking empire that rivalled the Ottomans in the west and the Mughals of India. Most importantly, it was Ismail who was responsible for the coupling of Iranian and Shia identities that is familiar to us today. He declared Twelver Shia as the official religion of the empire, and invited Shia scholars from all over the Islamic world to his court.[12]

Like other radical revolutionary regimes in history, the Safavids sought to revise prevailing notions of time and space. For some, this meant a return to the example of the Prophet, who used to pray due south. This approach was championed by one of the highest religious authorities of the new regime, ʿAlī ibn Ḥusayn al-Karakī (d. 1534). Al-Karakī, originally from the mountains of Lebanon, was among the Shia scholars flocking to the court of Shah Ismail in the early sixteenth century. As a fervent revolutionary, al-Karakī argued that, as the Prophet used to pray due south, all believers in all places, including Iran, should pray with 'Capricorn between their shoulders'.[13] Al-Karakī was adopting the example of the Prophet as well as the practice in his own native land of the mountains of Lebanon, where Mecca was indeed due south. At the same time, he was drawing on the tools of folk astronomy familiar to the lay person and disregarding any mathematical calculation of the *qibla* direction. Al-Karakī convinced the Safavid shah that the revolution would be complete only if the prayer niches of each and every mosque in Safavid Iran were realigned.[14]

Al-Karakī was followed by other traditionalist Shia scholars who sought guidance in the sayings of the Shia imams in the first centuries of Islam, and were sceptical about the mathematical sciences. For these traditionalist scholars, the science of astronomy was essentially speculative and had no foundation in scripture. Their claims had more than a grain of truth: the values given to coordinates, especially of longitude, were based on very flimsy foundations and subject to very wide disagreement between astronomers and different astronomical tables. Traditionalists also pointed out that the widely accepted assumption of a spherical Earth among astronomers had never been proven.[15] Indeed, a minority of medieval Muslim religious scholars argued that the Earth was flat, partly on the basis of the fact that the sun appears larger at sunset and sunrise and smaller at high noon.[16]

But over the next few decades al-Karakī's radical traditionalism gave way to an equally radical rationalist strand of Shia scholars, known as the *uṣūlīs*, who promoted the duty of a constant human quest to determine God's will, and coupled it with an active interest in mathematics and philosophy. In the following generation, one of the prominent representatives of the *uṣūlī* trend, the leading Shia scholar Ḥusayn ibn ʿAbd al-Ṣamad (d. 1576), argued that directing Iranians to pray due south not only was incorrect but indicated blind imitation and narrow-mindedness. In a treatise devoted to the determination of the prayer direction in Iran, Ḥusayn ibn ʿAbd al-Ṣamad demonstrated, on the basis of mathematical and astronomical proofs, that all regions of Iran lie further to the east in relation to Mecca, on a different longitude, and therefore praying due south diverted Iranians from the correct path. In Mashhad, for example, one should pray to the south-west, at 225 degrees, and the values are different for every town, depending on its geographical location.[17] The debate was not just about the practical point of the determination of the *qibla*. Ḥusayn ibn ʿAbd al-Ṣamad found fault with all blind reliance on the opinions of the scholars of the past, and called for every generation to continually investigate God's will. One could not rely on the opinions of dead jurists; qualified scholars had to find the truth for themselves.[18]

The matter of the determination of the *qibla* thus became a symbolic battle line between the two camps, the rationalists and the traditionalists, especially after Shah ʿAbbās I (1588–1629) moved the Safavid capital to Isfahan. The rationalist *uṣūlīs*, now favoured by the court, made Isfahan their home, where they created a unique intellectual milieu. The *uṣūlī* scholars of Isfahan were as much philosophers and scientists as they were jurists and theologians. Ḥusayn ibn ʿAbd al-Ṣamad's son, Shaykh al-Islam Bahā al-Dīn al-ʿAmilī (d. 1620 or 1621), was not only the city's supreme religious authority, but also a mathematician and a defender of the use of the astrolabe. As a skilled engineer, he was in charge of much of the rapid expansion of Isfahan into one of the world's largest cities, designing both public squares and residential areas. He also oversaw the construction of a sundial inside the Sulaymaniyya Madrasa and, we are told, performed the calculations that ensured that all new mosques in the city faced in the correct direction of prayer.

It was in this unique religious milieu of Safavid Isfahan that the production of astrolabes peaked, around 1640, before Isfahan was sacked by Afghan troops

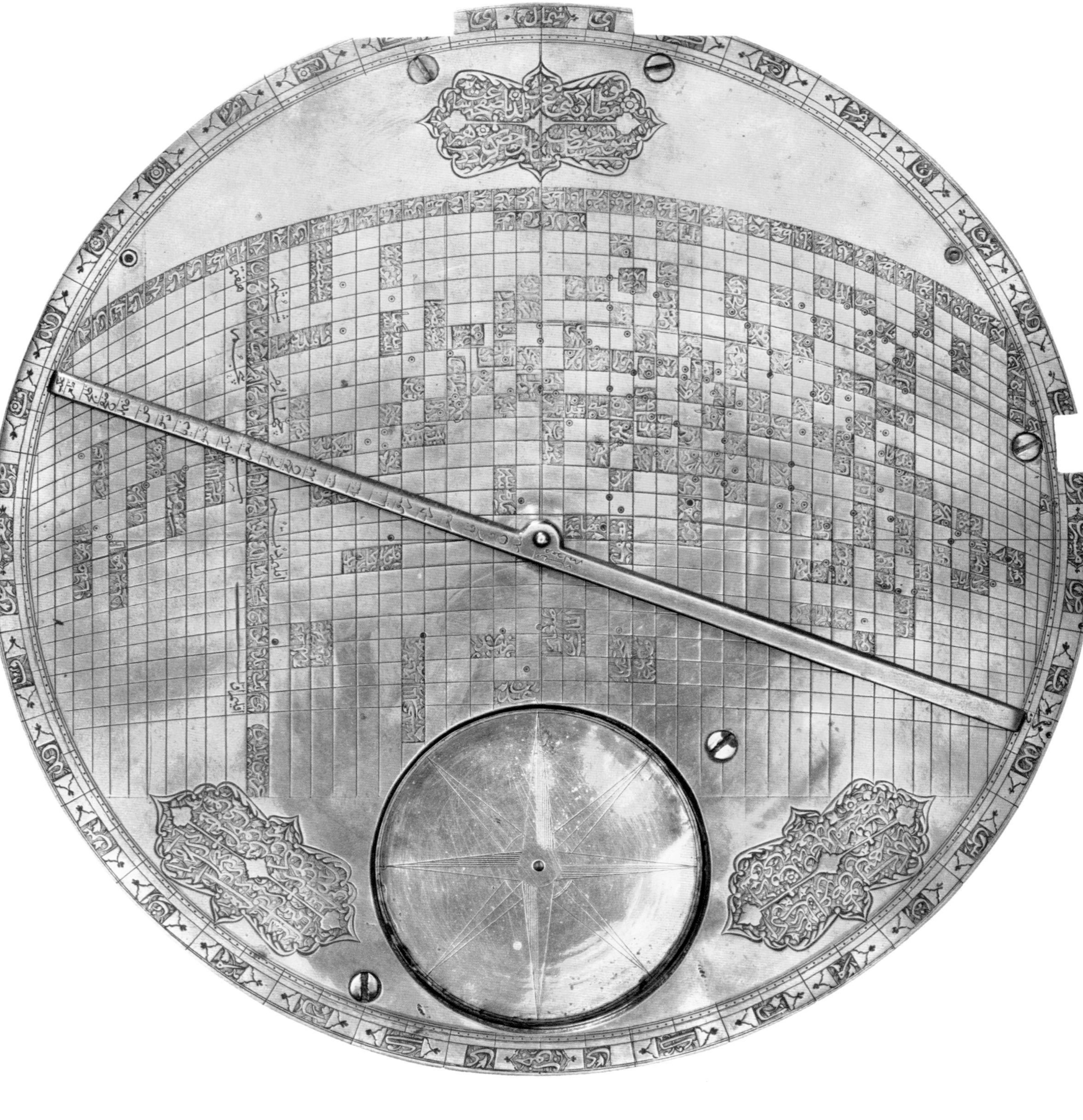

World map on brass for finding the direction to Mecca. Iran, 17th century. © The al-Sabah Collection, Dar al-Athar al-Islamiyyah, Kuwait, LNS 1106 M v4.

in 1722. For the *uṣūlīs*, these instruments symbolized the marriage of faith and science, a combination of piety and constant rational enquiry. The instruments enabled the shah himself and other members of the elite to fulfil God's commands with greater precision; each of these metal constructions carried an individual social, political and intellectual statement. As the French traveller Jean Chardin noted in 1666, everyone he met in court in Isfahan carried his own astrolabe, and 'guarded it like a jewel'.[19] At the same time, the instruments underscored the dynasty's claims to the pre-Islamic tradition of divine kingship. It was at this time that the Safavids started promoting and endorsing the Nowruz festival, the 'Imperial' (rather than Islamic) New Year, celebrated on 21 March. The key moment of these celebrations was when the chief imperial astronomer came to the palace and, after consulting his astrolabe, announced the precise moment of the equinox. Spring and creation would recommence according to the reading of the indications on the instrument.

In addition to astrolabes, Safavid Isfahan also saw the production of new devices for finding the *qibla*, and the most remarkable innovation was the production of world maps with Mecca at their centre, which came to the attention of modern scholars only in the late 1980s.[20] One of three known surviving artefacts of this kind is now housed in the Dar al-Athar al-Islamiyyah in Kuwait (opposite). Mecca is shown at the centre of the disc, at the point where the ruler is attached to the metal disc, and on a grid of latitude and longitude lines, mathematically plotted at 77° 10' east and 21° 40' north. The straight lines of longitude stretch 48 degrees west and east of Mecca, equally spaced at 2 degrees from each other; they reach up to the western tip of North Africa on the one side and to the extremity of China on the other. The curved latitudinal lines start at 10 degrees north at the bottom of the grid, the latitude of Aden, and go up to 50 degrees north, the latitude of the land of the Bulghars (present-day Bulgaria). The names of 140 cities and regions of the Islamic world are engraved into the squares formed by the grid, and each label lies next to a dot that marks the precise location on the grid.

The function of the ruler, revolving around the axis of Mecca, is to show both the direction of Mecca and its distance from any locality. Once the ruler is aligned with any of the dots, the direction to Mecca can be read from the finely executed scale along the circumference of the instrument. The scale markings on the ruler itself show the distance between Mecca and the chosen locality, measured in parsangs, each parsang being equal to the distance covered by a day's walk. The text in the cartouche on the bottom right of the disc gives instructions in Persian about the use of the ruler: 'When you put the diametrical ruler at the latitude and longitude of any city, you will be shown the direction and distance of the *qibla*.'[21]

The mathematical sophistication undergirding this instrument is remarkable, even if the actual plotting of localities is riddled with errors. Historians of science widely disagree (as they often do) about the exact method by which this device was constructed. As is evident, the projection used here is not the simple orthogonal projection typical of early medieval attempts to construct plotted world maps (discussed in Chapter 1). The curved latitudinal lines are required by the function of the map, since finding the correct direction and distance to the Kaaba has to take the curvature of the Earth into account. The calculation of the distance to the Kaaba from each locality is likely grounded in spherical trigonometry, based on the application of sine values to the Earth's surface. In principle, the instrument allowed the user to find the direction and distance to Mecca from any point in the Islamic world and beyond, as long as the coordinates of the user's location were known.

This Mecca-centred, mathematically plotted metal world map epitomizes the Safavid merging of astronomy and piety, of 'science in the service of Islam'.[22] Apart from the map itself, the recess at the bottom of the disc originally housed a magnetic compass, and a sundial, probably of a European design, was attached to the instrument. The Persian text in the two other cartouches explains how the sundial and the compass should be used. This rather heavy instrument would not have been taken on the road: with a radius of 22.5 centimetres, it was more of a display piece. But it is not particularly ornamental and was not intended for the shah himself. Two other specimens of this prototype have surfaced in the last two decades. The maker, who identified himself in one of the two other instruments as Muḥammad Ḥusayn, had clearly hoped for more than one customer.

By the early eighteenth century, instruments that combined a *qibla* indicator, a sundial and a compass became a popular commodity in Safavid Isfahan. Compared to the fancy astrolabes, these pious gadgets were easier to use and probably much cheaper. They were also increasingly associated with specifically Shia forms of devotion. The rectangular instrument shown here (opposite), held in the History of Science Museum in Oxford, may have been made by ʿAbd al-Aʾimmah, who is known to us by his signatures on several astrolabes he made for wealthy patrons in the Safavid court. But this is a much simpler device, which may have been used in a public space, probably a mosque or a square.

Rectangular pin-gnomon and *qibla* indicator from Iran, 18th century. © History of Science Museum, University of Oxford, 48472.

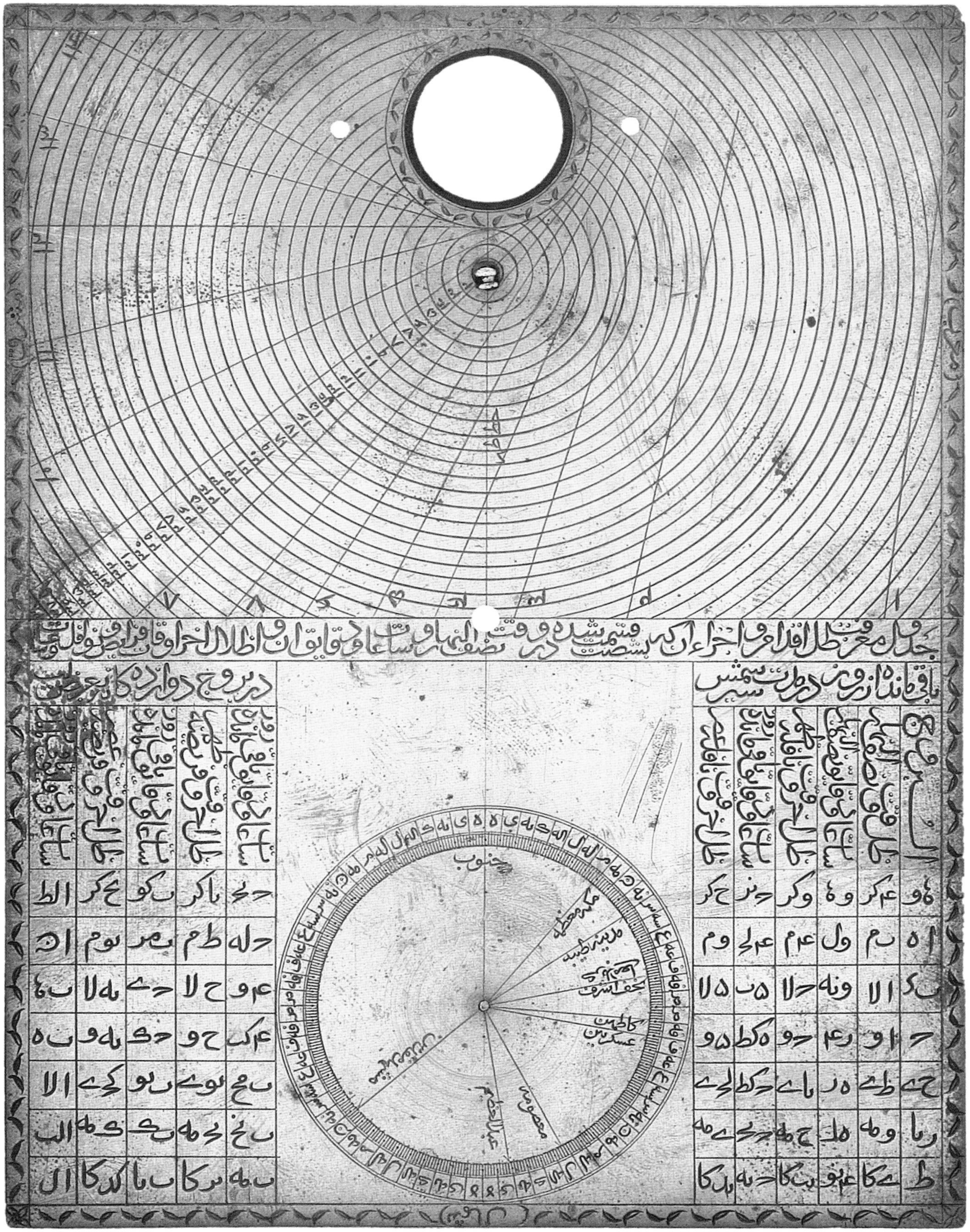
جنوب
مکه معظمه
مدینه طیبه
نجف اشرف
کاظمین
عسکریین

شمال
مدینه طیبه
مکه معظمه
بحرین
بغداد
موصل
همدان
کاشان
اصفهان
شیراز
اسکندریه
مصر
بیت المقدس
عسقلان
طبریه
بعلبک

Pocket *qibla* indicator and horizontal pin-gnomon from Iran, probably 19th century.

The upper half consists of a sundial; the lines for the hours can be seen, but the gnomon (the stick projecting the shadow) is missing. The disc in the lower part of the instrument locates Isfahan at the central point, with south at the top. The direction of Mecca is shown at 40 degrees south-west, equivalent to the standard 220-degree direction of prayer from Isfahan. The disc also shows the directions of other holy shrines, with most of them the burial sites of Shia imams. For example, the line indicating the direction of the sacred city of Mashhad in northern Iran is just opposite the line for the direction of Mecca; a user in Isfahan could also find here the direction of Najaf in southern Iraq, and of al-Maʿṣūmah, referring to the tomb of the Prophet's daughter Fāṭimah in the city of Qum. The sacred geography inscribed here is both universally Islamic and specifically Shia.

The production of simple *qibla* indicators continued even after the fall of the Safavids. This lovely small object was probably made as late as the nineteenth century (opposite). Mecca is back at the centre of the disc, which is oriented towards the north. Important cities and regions of the Islamic world are shown in relation to Mecca, and an attempt has been made to preserve not only the direction of the *qibla* but also the distance from it. In a way, this is a simplified miniature of the Mecca-centred world map discussed above, and it is likely that it also had a rotating ruler, which is now missing. In the west, the furthest city shown is Alexandria; the northernmost cities shown are in Anatolia (but do not include Istanbul). In the east, the coverage of Iranian cities is much denser, with Isfahan shown in a north-eastern direction from Mecca, midway between the centre of the circle and the outer boundaries of the Persian-speaking world. A simple sundial and a glass box for a magnetic compass used to occupy the lower half of the disc; both the gnomon and the compass itself are lost. The back of the instrument has a table, written in tiny letters, with the *qibla* directions of the thirty-two Muslim cities.

The owner of this instrument had the sacred world of Islam in his pocket.[23] In a period when mechanical pocket watches imported from Europe were becoming fashionable, this instrument relied on methods sanctioned by Islamic tradition. The sundial provided a sufficiently accurate measurement, enough for knowing the prayer times. The compass was sufficient for finding the direction of prayer and, even if it were lost or malfunctioned, the sundial could also be used for basic orientation. The map itself, covering the central lands of Islam, from Iran to Egypt, could allow merchants to direct their prayers towards Mecca at any conceivable stop during their travels. Beyond its practical use, this was also an object of devotion, the owner reminding themselves and others of the presence of Allah. And, like the astrolabe made for ʿAbbās II, the instrument allowed the owner to read the measurements for themselves; it enabled them to participate in a scientific, rational enquiry about the direction of prayer, and by implication about God's will from His creation.

Mecca tile from Iznik or Kütahya, in present-day Turkey, *c.*1650. © The Trustees of the British Museum, 2009. 6039. 1.

Islam and Maps

Over the history of Islam, the visual projection of devotion to the Kaaba has taken many forms. Sometimes this was done through maps or diagrams that showed the Kaaba at the centre of a circle made of eight or more sectors, with each region of the known world facing one wall or one corner of the Kaaba. Other Muslims sought to visualize the direction to Mecca using direct lines and precise calculations. The difference between these options was partly due to the technology and knowledge available – the compass allowed much more precision in orienting one's prayer, and advances in spherical trigonometry and in the observation of celestial bodies opened the way to precise mathematical solutions. But variations in cartographic representation also came down to different approaches to the relationship between Islam and science.

In Islamic legal literature, the problem of determining the direction of prayer is seen as paradigmatic of all questions that require independent enquiry on the part of the believer. The maps for finding the direction to Mecca reflect opposing approaches to the role of that independent enquiry. Most medieval charts only gave approximate directions and left room for ambiguity and flexibility. There was always more than one answer to the matter of determining the *qibla*. The Safavid instruments, associated with a politically active and philosophically oriented strand of Shia thought, wished to eliminate ambiguity and plurality, and to replace it with mathematical precision. Since then, more and more modern Muslims have wished to harness the certainty of science in the service of their faith. But this approach is not without its problems. Is the prayer of a person who finds the direction to Mecca by a global positioning system more valuable or meritorious than the prayer of a person who relies on an estimate? Are they a better Muslim?

Medieval *qibla* maps had other functions. For example, they made the Kaaba seem closer, within reach. The earlier, abstract diagrams in particular showed the details of the rectangular structure, the location of the Black Stone in the eastern corner and other holy sites in the Holy Sanctuary in Mecca. For the person

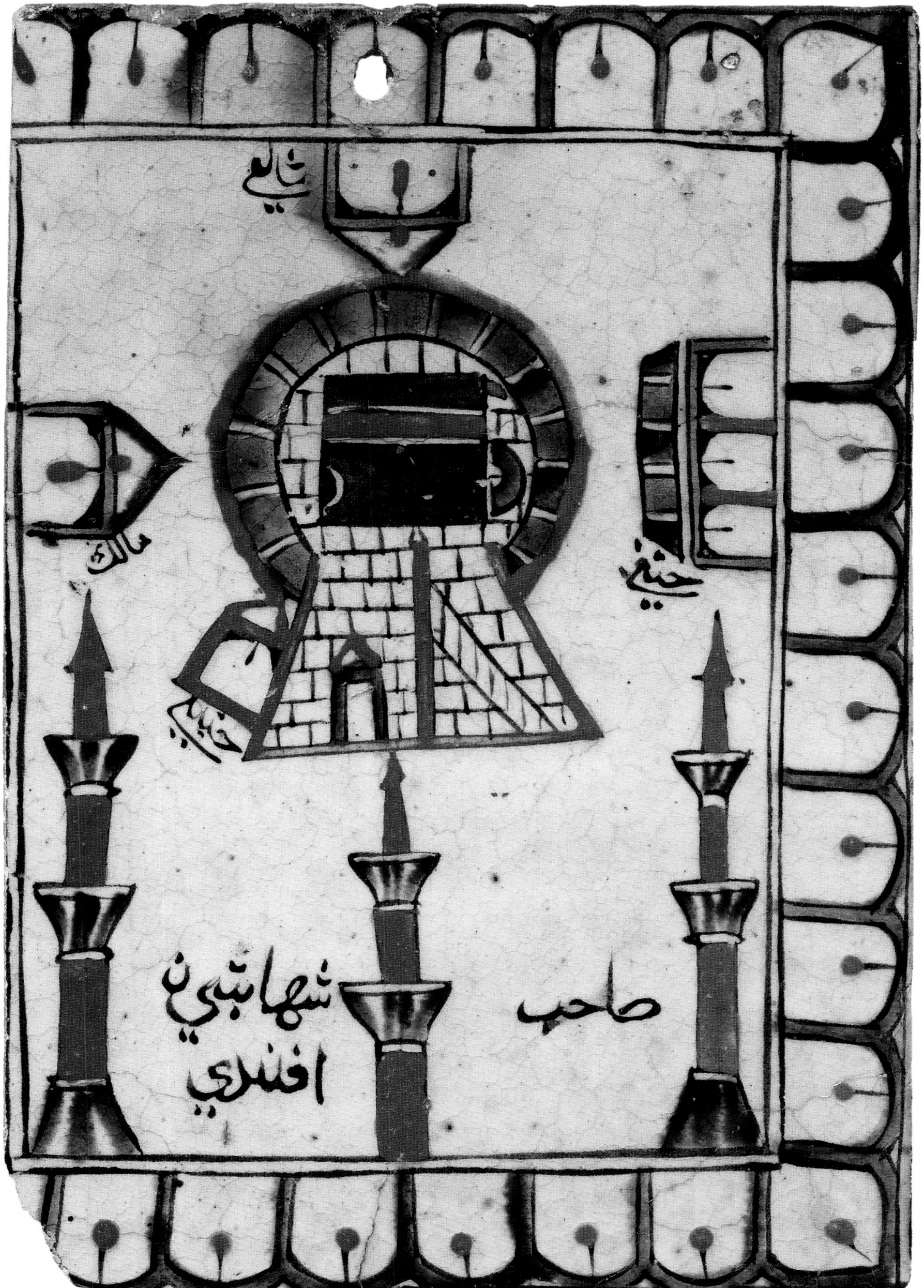
شافعي
مالك
حنفي
حنبلي
صاحب
شهابتين
افندي

praying in faraway lands, they made the orientation towards Mecca concrete and meaningful, regardless of mathematical accuracy. Such visualizations had much in common with the two-dimensional view of the Holy Sanctuary that illustrated pilgrimage guides and the cherished certificates carried back by pilgrims who had performed the hajj. By the seventeenth century, Ottoman artists had translated this two-dimensional view into tile art (p. 177). Beautiful tiles showing the Holy Sanctuary, glazed in vibrant colours, were produced at the ceramic centres of Kütahya and Iznik. The earliest surviving example is on the eastern wall of Hagia Sofia mosque in Istanbul. It is dated 1642, roughly contemporaneous with the astrolabe made in Isfahan for Shah ʿAbbās.[24]

A tradition attributed to the Prophet holds that Mecca is the navel of the Earth.[25] But Muslims did not literally take Mecca to be the centre of the world, not even in the *qibla* diagrams, quadrants, tables and indicators discussed in this chapter. The direction of Mecca was important, but it didn't fundamentally shape prevalent geographical concepts. Al-Iṣṭakhrī opened his set of maps of the world of Islam with a depiction of the Arabian Peninsula. A Fatimid caliph indicated Mecca and Medina prominently on his exquisite world map, and on the mysterious rectangular world map in the *Book of Curiosities* Mecca is marked by a distinctive yellow horseshoe. But the highlighting of Mecca is at the margins of what these map-makers were trying to achieve in their maps, and is completely absent from al-Idrīsī's grid and from the maritime charts of Pīrī Reʾīs.

In truth, Mecca and the Kaaba were not central to most pre-modern Islamic visions of the world, which tended to separate theological narrative and cartographic representation. The maps produced in medieval Islamic societies projected political power, the unity of the Islamic world or the universalism of scientific enquiry. Very rarely, if at all, did they project images of salvation and theological narratives, in stark contrast to most medieval European maps, which showed paradise at the eastern edge of the sphere. This is possibly because Muslim theology never imagined the Kaaba to be the physical abode of God. While Jews prayed towards the Temple in Jerusalem because it used to house God's presence, the Qur'an and later commentators saw the injunction to pray towards Mecca as an act of commemoration and as a marker of group identity.[26]

Despite common wisdom, the southern orientation of many Islamic maps has nothing to do with Mecca's sanctity. There is in fact no uniform orientation to Islamic maps, and we have seen various orientations throughout our journey through *Islamic Maps*. South is at the top of most medieval Islamic world maps but this is most probably simply a convention. This common misconception regarding the orientation of Islamic maps is a reminder that it is neither necessary nor desirable to reduce Islamic maps to the religious beliefs of their makers. As this book has shown, these maps – as works of art, science, political propaganda and as objects of devotion – uniquely reflect the incredible richness, plurality and wide horizons of the societies that produced them.

Timeline

170 CE	Death of the Greek geographer Ptolemy of Alexandria
622	Hijra: migration of the Prophet Muhammad from Mecca to Medina
	Beginning of the Islamic calendar
632	Death of the Prophet Muhammad in Medina
632–656	Arab Muslim conquest of the Middle East
661	Establishment of the Umayyad dynasty in Damascus
711	Muslim conquest of Spain
750	Establishment of the Abbasid dynasty
762	Foundation of Baghdad as the Abbasid capital
813–833	Reign of the Abbasid caliph al-Maʾmūn
827	Start of Muslim rule in Sicily
*c.*847	Death of al-Khwārazmī
909	The establishment of the Fatimid Shia dynasty in North Africa
*c.*950	Death of al-Iṣṭakhrī
971	Foundation of Cairo as the Fatimid capital
1068–1071	Norman conquest of Sicily
1099	Crusader conquest of Jerusalem
1130–1154	Reign of Roger II, king of Sicily
1171	End of the Fatimid dynasty in Cairo
1258	End of the Abbasid dynasty in Baghdad
1260	Rise of the Mamluk dynasty in Egypt and Syria
1453	Ottoman conquest of Constantinople/Istanbul
1492	Fall of Granada: end of Islamic rule in Spain
1501	Establishment of the Safavid Shia dynasty in Iran
1517	Ottoman conquest of Syria and Egypt
	End of the Mamluk dynasty
1554	Death of Pīrī Reʾīs
1632–1666	Reign of Shāh ʿAbbās II of Iran
1765	End of Safavid rule in Iran

Notes

Introduction

1 Recent books include Karen Pinto, *Medieval Islamic Maps: An Exploration*, University of Chicago Press, Chicago, IL, and London, 2016; Tarek Kahlaoui, *Creating the Mediterranean: Maps and the Islamic Imagination*, Brill, Leiden and Boston, 2018; Zayde Antrim, *Mapping the Middle East*, Reaktion Books, London, 2018; and Yossef Rapoport and Emilie Savage-Smith, *Lost Maps of the Caliphs: Drawing the World in Eleventh-Century Cairo*, University of Chicago Press, Chicago, IL, and London; Bodleian Library, Oxford, 2018. All these volumes build on the foundations laid in the standard introduction to Islamic maps, J.B. Harley and David Woodward (eds), *History of Cartography*, vol. 2, book 1, *Cartography in the Traditional Islamic and South Asian Societies*, University of Chicago Press, Chicago, IL, and London, 1992, which Chicago University Press have made available in full online at www.press.uchicago.edu/ucp/books/book/chicago/H/bo3625863.html, accessed 18 February 2019.

2 Jerry Brotton, *A History of the World in Twelve Maps*, Penguin Books, London, 2013, p. 14.

3 J.B. Harley and David Woodward (eds), *History of Cartography*, vol. 1, *Cartography in Prehistoric, Ancient, and Medieval Europe and the Mediterranean*, University of Chicago Press, Chicago, IL, and London, 1987, p. xvi.

4 ʿUbayd Allāh ibn ʿAbd Allāh ibn Khurradādhbih, *[Kitāb] al-Masālik wa-al-Mamālik*, ed. Michael Jan de Goeje, Brill, Leiden, 1889, pp. 182–3, cited in Travis Zadeh, 'Of Mummies, Poets, and Water Nymphs: Tracing the Codicological Limits of Ibn Khurradādhbih's Geography', in Monique Bernards (ed.), *ʿAbbasid Studies IV: Occasional Papers of the School of ʿAbbasid Studies (Leuven, July 5–July 9, 2010)*, Gibb Memorial Trust, Warminster, 2013, pp. 8–75. On the magical powers of maps, see Brotton, *A History of the World in Twelve Maps*, p. 3.

Chapter 1

1 The Arabic text has been published as al-Khwārazmī, *Das Kitāb Ṣūrat al-Arḍ des Abū Ǧaʿfar Muḥammad ibn Mūsā al-Khuwārizmī*, ed. Hans von Mžik, Bibliothek Arabischer Historiker und Geographen 3, Otto Harrassowitz, Leipzig, 1926. The work as a whole has not been translated into a European language. The longitude and latitude tables that take up most of the treatise are summarized in E.S. Kennedy and M.H. Kennedy, *Geographical Coordinates of Localities from Islamic Sources*, Institut für Geschichte der Arabisch-Islamischen Wissenschaften an der Johann Wolfgang Goethe-Universität, Frankfurt am Main, 1987.

2 On the depiction of the sources of the Nile in classical and Islamic geographical sources, see Robin Seignobos, 'L'origine occidentale du Nīl dans le géographie latine et arabe avant le XIVe siècle', in Nathalie Bouloux, Anca Dan and Georges Tolias (eds), *Orbis Disciplinae: hommages en l'honneur de Patrick Gautier-Dalché*, Brepols, Turnhout, 2017, pp. 371–94. On Ptolemy's depiction of the sources of the Nile and its reception by medieval European authors, see Patrick Gautier-Dalché, *La géographie de Ptolémée en Occident (IVe–XVIe siècle)*, Brepols, Turnhout, 2009, p. 58.

3 ʿAlī ibn al-Ḥusayn al-Masʿūdī, *Kitāb al-Tanbīh wa-al-Ishrāf*, ed. Michael Jan de Goeje, Brill, Leiden, 1894, p. 33. My translation here differs slightly from the version in G.R. Tibbetts, 'The Beginnings of a Cartographic Tradition', in Harley and Woodward (eds), *History of Cartography*, vol. 2, book 1, *Cartography in the Traditional Islamic and South Asian Societies*, p. 96.

4 Fuat Sezgin, *Mathematische Géographie und Kartographie im Islam und ihr Fortleben im Abendland: Kartenband*, Geschichte der Arabischen Schrifttums 12, Institut für Geschichte der Arabisch-Islamischen Wissenschaften an der Johann Wolfgang Goethe-Universität, Frankfurt am Main, 2000, p. 4. See also Jean-Charles Ducène, 'L'Afrique dans les mappemondes circulaires arabes médiévales: typologie d'un représentation', in Robin Seignobos and Vincent Hiribarren (eds), 'Cartographier l'Afrique: construction, transmission et circulation des savoirs géographiques du Moyen Âge au XIX siècle', special issue, *Cartes & Géomatique: Revue du Comité Français de Cartographie*, no. 210, 2011, pp. 19–36, at p. 31, fig. 1.

5 See al-Khwārazmī, *Das Kitāb Ṣūrat al-Arḍ*, p. 139, line 4 ('a city that has no name on the map') and 77, line 9 ('other [rivers?] that are not named on the map'). See also Tibbetts, 'The Beginnings of a Cartographic Tradition', p. 100.

6 For further discussion of this map of the Nile, see Rapoport and Savage-Smith, *Lost Maps of the Caliphs*, pp. 101–24.

7 For an analysis by Raman spectroscopy of the pigments used in the manuscripts, see Tracey D. Chaplin, Robin J.H. Clark, Alison McKay and Sabina Pugh, 'Raman Spectroscopic Analysis of Selected Astronomical and Cartographic Folios from the Early 13th-Century Islamic Book of Curiosities of the Sciences and Marvels for the Eyes', *Journal of Raman Spectroscopy*, vol. 37, 2006, pp. 865–77.

8 Kennedy and Kennedy, *Geographical Coordinates of Localities from Islamic Sources*, p. 377.

9 Gerald R. Tibbetts, 'Later Cartographic Developments', in Harley and Woodward (eds), *History of Cartography*, vol. 2, book 1, *Cartography in the Traditional Islamic and South Asian Societies*, p. 153. Fuat Sezgin, the first to recognize the significance of this map, has argued that it is a copy of a plotted map al-Khwārazmī prepared for the caliph al-Maʾmūn (Sezgin, *Mathematical Geography*

and Cartography in Islam and their Continuation in the Occident, 3 vols, English translation of vols 10–12 of *Geschichte des Arabischen Schrifttums*, Institute for the History of Arabic-Islamic Science, Frankfurt am Main, 2000–7).

10 Yossef Rapoport and Emilie Savage-Smith (eds and trans), *An Eleventh-Century Egyptian Guide to the Universe: The 'Book of Curiosities'*, Islamic Philosophy, Theology and Science, Texts and Studies 87, Brill, Leiden, 2014, pp. 442–3.

Chapter 2

1 This is the title of al-Iṣṭakhrī's work in its late medieval Persian translations. The original Arabic work was probably called *Maps of the Regions of the World* (*Ṣuwar al-Aqālīm*). See Jean-Charles Ducène, 'Quel est le titre véritable de l'ouvrage géographique d'al-Iṣṭaḫrī?', *Acta Orientalia Belgica*, vol. 19, 2006, pp. 99–108.

2 The maps made by al-Iṣṭakhrī are discussed in more detail in several major recent publications: Pinto, *Medieval Islamic Maps*; Antrim, *Mapping the Middle East*; and Nadja Danilenko's PhD dissertation, 'Picturing the Islamicate World in the Tenth Century: The Story of al-Iṣṭaḫrī's *Book of Routes and Realms*', Freie Universität Berlin, 2018. I am grateful to Nadja Danilenko for going through a draft of this chapter and making numerous suggestions and corrections.

3 Emilie Savage-Smith, 'Memory and Maps', in Farhad Daftary and Josef W. Meri (eds), *Culture and Memory in Medieval Islam: Essays in Honour of Wilferd Madelung*, I.B. Tauris, London, 2003, pp. 109–27, figs 1–4.

4 Zayde Antrim, *Routes and Realms: The Power of Place in the Early Islamic World*, Oxford University Press, Oxford, 2012.

5 Al-Iṣṭakhrī, *Kitāb al-Masālik wa-al-Mamālik = Viae regnorum: descriptio ditionis moslemicae*, ed. Michael Jan de Goeje, Bibliotheca Geographorum Arabicorum 1, Brill, Leiden, 1873; repr. 1967, p. 3.

6 For a recent reassessment of the biography of Ibn Ḥawqal and the surviving manuscripts of his work, see Chafik Benchekroun, 'Requiem pour Ibn Hawqal: sur l'hypothèse de l'espion fatimide', *Journal Asiatique*, vol. 304, no. 2, 2016, pp. 193–211.

7 Antrim, *Routes and Realms*, pp. 117–19.

8 Al-Muqaddasī, *Aḥsan al-Taqāsīm fī Maʿrifat al-Aqālīm*, ed. Michael Jan de Goeje, Bibliotheca Geographorum Arabicorum 3, Leiden, Brill, 1877, p. 10 (my translation); translated as *The Best Divisions for Knowledge of the Regions: A Translation of 'Ahsan al-Taqāsīm fī Maʿrifat al-Aqālīm'* by Basil Anthony Collins, Garnet Publishing/Centre for Muslim Contribution to Civilization, Reading, 1994.

9 Al-Muqaddasī, *Aḥsan al-Taqāsīm*, p. 11.

10 Ibn Ḥawqal, *Kitāb Ṣūrat al-Arḍ*, ed. J.H. Kramers, 2nd edn, 2 vols, Bibliotheca Geographorum Arabicorum 2, Brill, Leiden, 1938–9, pp. 2–3 (my translation). For a somewhat different interpretation of this passage, see Antrim, *Routes and Realms*, pp. 111–12, 114 (arguing for a broader audience).

11 See Danilenko, 'Picturing the Islamicate World in the Tenth Century'.

12 See the Indian Ocean map in the copy preserved in the British Library, Or. 1587, fol. 39r, which is reproduced in Gerald R. Tibbetts, 'The Balkhī School of Geographers', in Harley and Woodward (eds), *History of Cartography*, vol. 2, book 1, *Cartography in the Traditional Islamic and South Asian Societies*, p. 127, fig. 5.23. Tibbetts had identified the manuscript as originating in India, a mistake rectified by Nadja Danilenko.

13 Pinto, *Medieval Islamic Maps*, pp. 219–78.

14 Pinto argues that the fifteenth-century copyists introduced changes to the labels on the world map to reflect Ottoman territorial claims as the heirs of the Byzantine empire (*Medieval Islamic Maps*, pp. 271–8). In my view, the changes are minimal, and the copy is largely very faithful to al-Iṣṭakhrī's tenth-century world view.

Chapter 3

1 On the discovery of the *Book of Curiosities* and its contributions to the history of cartography and to the history of global communications at the turn of the previous millennium, see Rapoport and Savage-Smith, *Lost Maps of the Caliphs*. The text of the treatise has been edited and translated by Rapoport and Savage-Smith as *An Eleventh-Century Egyptian Guide to the Universe*.

2 Farhad Daftary, *The Ismaʿilis: Their History and Doctrines*, 2nd edn, Cambridge University Press, Cambridge, 2007.

3 Ibn Ḥawqal, *Kitāb Ṣūrat al-Arḍ*, p. 71 (my translation).

4 Muḥammad ibn Muḥammad al-Idrīsī, *Nuzhat al-Mushtāq fī Ikhtirāq al-Āfāq: opus geographicum, sive 'Liber ad eorum delectationem qui terras peragrare studeant'*, 9 parts in 2 vols, ed. Alessio Bombaci, Umberto Rizzitano, Roberto Rubinacci and Laura Veccia Vaglieri, Istituto Universitario Orientale di Napoli/Istituto Italiano per il Medio ed Estremo Oriente, Naples, 1970–6, vol. 1, p. 282.

5 This fascinating and often quite explicit ethnography of the Arabia Peninsula in the Late Middle Ages has recently been translated into English by G. Rex-Smith as *A Traveller in Thirteenth-Century Arabia: Ibn al-Mujāwir's 'Tārīkh al-Mustabṣir'*, Routledge, London, 2017.

6 *Irtifāʿ al-dawlah al-Muʾayyadīyah: jibāyat bilād al-Yaman fī ʿahd al-Sulṭān al-Malik al-Muʾayyad Dāwūd ibn Yūsuf al-Rasūlī, al-mutawaffá sanat 721 H/1321 M*, ed. Muḥammad ʿAbd al-Raḥīm Jāzim, Centre Français d'Archeologie et de Sciences Sociales de Sanaa, Sanaa, 2008.

Chapter 4

1 Tariq Ali, *A Sultan in Palermo*, Verso, London and New York, 2006.

2 Muḥammad ibn Muḥammad al-Idrīsī, *La première géographie de l'Occident*, trans. Amédée Jaubert, rev. Henri Bresc and Annlies Nef, Flammarion, Paris, 1999.

3 Brotton, *A History of the World in Twelve Maps*, pp. 54–81.

4 S. Maqbul Ahmad, 'Cartography of al-

Sharīf al-Idrīsī', in Harley and Woodward (eds), *History of Cartography*, vol. 2, book 1, *Cartography in the Traditional Islamic and South Asian Societies*, pp. 156–74.

5 An excellent summary of recent scholarship is presented in Jean-Charles Ducène, 'Al-Idrīsī, Abū ʿAbdallāh', in *Encyclopaedia of Islam*, 3rd edn, ed. Kate Fleet, Gudrun Krämer, Denis Matringe, John Nawas and Everett Rowson, https://referenceworks.brillonline.com/browse/encyclopaedia-of-islam-3, accessed 18 February 2019. This entry also lists all the translations and modern studies of al-Idrīsī.

6 This text was first identified by Alloua Amara and Annlies Nef, 'Al-Idrīsī et les Hammūdides de Sicile: nouvelles données biographiques sur l'auteur du *Livre de Roger*', *Arabica*, vol. 48, 2001, pp. 121–7. See also Wael Abu-ʿUksa, 'Lives of Frankish Princes from al-Ṣafadī's biographical dictionary, *al-Wāfī bil-Wafayāt*', *Mediterranean Historical Review*, vol. 32, no. 1, 2017, pp. 83–104.

7 Al-Idrīsī, *Nuzhat*, p. 5 (my translation).

8 Abu-ʿUksa, 'Lives of Frankish Princes'.

9 Al-Idrīsī, *Nuzhat*, p. 6 (my translation).

10 Al-Maqrīzī, *al-Mawāʿiẓ wa-al-Iʿtibār fī Dhikr al-Khiṭaṭ wa-al-Āthār*, 4 vols, ed. Ayman Fuʾad Sayyid, Muʾassasat al-Furqān li'l-Turāth al-Islāmī, London, 2002, vol. II, p. 305 (my translation).

11 The *Entertainment* contains references to events in 1157 or even 1158, such as Frederick Barbarossa's establishment of himself in Bourgogne. See al-Idrīsī, *La première géographie de l'Occident*, pp. 18, 40.

12 Nine sectional maps of one early copy of al-Idrīsī's work, dated to *c.*1300, indicate in the map labels numerical values for localities in the first clime, near the equator. Ducène is of the opinion that these coordinate values were introduced by al-Idrīsī himself, but they are not in the text of the treatise and may well have been added by a copyist. See Jean-Charles Ducène, 'Les coordonnées géographiques de la carte manuscrite d'al-Idrīsī (Paris, Bnf ar. 2221)', *Der Islam*, vol. 86, no. 2, 2009, pp. 271–85.

13 Brotton, *A History of the World in Twelve Maps*, p. 75.

14 Tarek Kahlaoui, 'The Maghrib's Medieval Mariners and Sea Maps', *Journal of Historical Sociology*, vol. 30, 2017, p. 47.

15 Brotton, *A History of the World in Twelve Maps*, p. 81.

16 For an authoritative discussion of this subject, see Jean-Charles Ducène, *L'Europe et les géographes arabes du Moyen Age (IXe-XVe siècle): 'La grande terre' et ses peoples: conceptualisation d'un espace ethnique et politique*, CNRS, Paris, 2018.

17 Jean-Charles Ducène, 'L'Europe dans le cartographie arabe médiévale', *Belgeo: Revue Belge de Géographie*, vols 3–4, 2008, pp. 261–2.

18 A.F.L. Beeston, 'Idrisi's Account of the British Isles', *Bulletin of the School of Oriental and African Studies*, vol. 13, no. 2, 1950, p. 278.

19 Kahlaoui, 'The Maghrib's Medieval Mariners and Sea Maps'.

20 Al-Idrīsī, *Nuzhat*, p. 13 (my translation).

Chapter 5

1 The best short introductions to Pīrī Reʾīs's life and work are still Svat Soucek, 'Islamic Charting in the Mediterranean', in Harley and Woodward (eds), *History of Cartography*, vol. 2, book 1, *Cartography in the Traditional Islamic and South Asian Societies*, pp. 263–87; and Soucek, 'Pīrī Reʾīs', in *Encyclopaedia of Islam*, 2nd edn, ed. P. Bearman, Th. Bianquis, C.E. Bosworth, E. van Donzel and W.P. Heinrichs, https://referenceworks.brillonline.com/browse/encyclopaedia-of-islam-2, accessed 25 May 2018. Another useful guide, especially to the manuscripts, is Mine Esiner Özen, *Pirî Reis and his Charts*, N. Refioğlu, Istanbul, 1998. There are several editions of Pīrī's world map and his *Book on Seafaring*. The most recent editions and translations are Fikret Sarıcaoğlu, *Pîrî Reîs in Dünya Haritası 1513 = The World Map of Pîrî Reîs 1513*, T.C. Kültür ve Turizm Bakanlığı, Ankara, 2014, and Pîrî Reis, *The Book of Bahriye*, ed. and trans. Bülent Özükan, Boyut Yayınları, Istanbul, 2013.

2 Kahlaoui, 'The Maghrib's Medieval Mariners and Sea Maps'.

3 Sarıcaoğlu, *The World Map of Pîrî Reîs*, p. 43. On the world map, see also Gregory C. McIntosh, *The Piri Reis Map of 1513*, University of Georgia Press, Athens, GA, 2000.

4 Sarıcaoğlu, *The World Map of Pîrî Reîs*, p. 88.

5 Ibid., pp. 97–8.

6 Ibid., p. 27.

7 Pîrî Reis, *The Book of Bahriye*, p. 14.

8 See Cristoforo Buondelmonti, *Description of the Aegean and Other Islands*, ed. and trans. Evelyn Edson, Italica Press, New York, 2018.

9 Pîrî Reis, *The Book of Bahriye*, p. 11.

10 Ibid., pp. 203–4.

11 On Maṭrāqci Naṣūh, see J.M. Rogers, 'Itineraries and Town Views in Ottoman Histories', in Harley and Woodward (eds), *History of Cartography*, vol. 2, book 1, *Cartography in the Traditional Islamic and South Asian Societies*, pp. 228–55.

12 Henghar Watenpaugh, *The Image of an Ottoman City: Imperial Architecture and Urban Experience in Aleppo in the 16th and 17th Centuries*, Brill, Boston, MA, 2004, p. 227.

13 Giancarlo Casale, *The Ottoman Age of Exploration*, Oxford University Press, Oxford and New York, 2010.

14 On the link between Pīrī's maps and imperial policy, see Pinar Emiralioğlu, *Geographical Knowledge and Imperial Culture in the Early Modern Ottoman Empire*, Ashgate, Burlington, VT, 2014.

Chapter 6

1 On Islamic astrolabes, see Emilie Savage-Smith, 'Celestial Mapping', in Harley and Woodward (eds), *History of Cartography*, vol. 2, book 1, *Cartography in the Traditional Islamic and South Asian Societies*, pp. 12–70, and the Astrolabe home page on the History of Science Museum website (www.mhs.ox.ac.uk/astrolabe, accessed 18 February 2019).

2 For clear instructions on the use of a *qibla* quadrant, see Emily Winterburn, 'Using an Astrolabe', http://muslimheritage.com/article/using-astrolabe, accessed 18 February 2019.

3 See, for example, eQibla (ver 1.0), developed by by Hamid Zarrabi-Zadeh (http://eqibla.com); and Qiblaway, which uses Google Maps software to calculate the *qibla* direction with the help of latitude and longitude values for the shortest distance (www.qiblaway.com).

4 For the full inscription, see the record for the Shah ʿAbbās II astrolabe on the History of Science Museum website: www.mhs.ox.ac.uk/astrolabe/catalogue/browseReport/Astrolabe_ID=260.html, accessed 18 February 2019.

5 Three different people signed their names on the instrument made for Shah ʿAbbās, reflecting the complexity of its production: the maker of the astrolabe was Muḥammad Muqīm from Yazd, the engraver was Faḍl Allāh from Sabzawār, and the court astronomer who oversaw the work was Muḥammad Shāfiʿ from Junābid.

6 The best short introductions to the history of maps showing the direction of prayer are David A. King, 'Makka: 4. As the Centre of the World', in *Encyclopaedia of Islam*, 2nd edn; David A. King and Richard P. Lorch, 'Qibla Charts, Qibla Maps, and Related Instruments', in Harley and Woodward (eds), *History of Cartography*, vol. 2, book 1, *Cartography in the Traditional Islamic and South Asian Societies*, pp. 189–205; and Antrim, *Mapping the Middle East*, pp. 53–7. David King has published widely on the subject, and much of the following is based on his numerous articles, collected as *In Synchrony with the Heavens: Studies in Astronomical Timekeeping and Instrumentation in Medieval Islamic Civilization*, 2 vols, Brill, Leiden, 2004. See also Mònica Rius, *La alqibla en al-andalus y al-Magrib al-Aqsà*, Universitat de Barcelona, Facultat de Filologia, Barcelona, 2000.

7 On the fluid orientation of early mosques, see David King, 'From Petra Back to Makka: From "Pibla" back to Qibla', www.muslimheritage.com/article/from-petra-back-to-makka, accessed 18 February 2019. The article is a critique of a provocative monograph which argues that Muslims originally prayed towards Petra in modern Jordan: see Dan Gibson, *Early Islamic Qiblas: A Survey of Mosques Built between 1AH/622 CE and 263 AH/876 CE*, Independent Scholars Press, Vancouver, BC, 2017.

8 See discussion of this scheme in David King, *World Maps for Finding the Direction and Distance of Mecca: Innovation and Tradition in Islamic Science*, Al-Furqān Islamic Heritage Foundation/Brill, London, Leiden and Boston, 1999, p. 52.

9 Ali Moussa, 'Mathematical Methods in Abū al-Wafāʾ's Almagest and the Qibla Determinations', *Arabic Sciences and Philosophy*, vol. 21, no. 1, 2011, pp. 1–56.

10 Al-Dimyāṭī, *Kitāb al-Tahdhīb fī Adillat al-Qibla wa-Naṣb al-Maḥārīb*, Bodleian Library, MS Marsh 592, fols 7b-21b.

11 Venetia Porter (ed.), *Hajj: Journey to the Heart of Islam*, British Museum Press, London, 2012, p. 64.

12 On the Safavids, see Andrew J. Newman, *Safavid Iran: Rebirth of a Persian Empire*, I.B. Tauris, London, 2006.

13 Devin J. Stewart, 'Notes on the Migration of ʿAmili Scholars to Safavid Iran', *Journal of Near Eastern Studies*, vol. 55, 1996, p. 98.

14 Andrew J. Newman, 'Towards a Reconsideration of the "Isfahān School of Philosophy": Shaykh Bahāʾī and the Role of the Safawid ʿ*Ulamā*ʾ', *Studia Iranica*, vol. 15, 1986, pp. 165–99.

15 Newman, 'Towards a Reconsideration of the "Isfahān School of Philosophy"'.

16 Abū Rashīd al-Nīsābūrī (d. *c*.460/1068), *al-Masāʾil fī al-Khilāf bayna al-Basrīyīn wa-al-Baghdādīyīn*, ed. Maʿn Ziyāda and Riḍwān al-Sayyid, Maʿhad al-Inmāʾ al-ʿArabī, Tripoli, 1979, pp. 100–4. I owe this reference to Omar Anchassi.

17 Afandī, ʿAbd Allāh ibn ʿĪsá (d. *c*.1718), *Rīyāḍ al-ʿulamāʾ wa-Ḥiyāḍ al-fuḍalāʾ*, ed. Aḥmad Ḥusaynī, Matḅaʿat al-Khayyām, Qum, 1980, vol. 2, p. 111.

18 Abisaab, *Converting Persia: Religion and Power in the Safavid Empire*, I.B. Tauris, London, 2004, pp. 33, 37, 51.

19 Chardin, *Voyages du chevalier Chardin en Perse, et autres lieux de l'Orient*, Amsterdam, 1735, vol. IV, p. 332, quoted in H. Winter, 'Persian Science in Safavid Iran', in P. Jackson and L. Lockhart (eds), *The Cambridge History of Iran*, Cambridge University Press, Cambridge, 1986, vol. 6, p. 595.

20 These instruments are the main subject of a major monograph by David A. King, who has brought them to scholarly attention in *World Maps for Finding the Direction and Distance of Mecca*.

21 King, *World Maps for Finding the Direction and Distance of Mecca*, p. 204.

22 See David A. King, 'Astronomy in the Service of Islam', lecture delivered at the al-Furqān Islamic Heritage Foundation, London, 7 March 2018, www.al-furqan.com/gallery/id/2668/filetype/video, accessed 18 February 2019.

23 I owe this observation to Silke Ackermann, director of the History of Science Museum in Oxford. The museum houses one of the world's largest collections of *qibla* indicators.

24 Porter (ed.), *Hajj*, p. 118.

25 Other traditions, which are in fact more common in early Islamic literature, consider either Baghdad or Jerusalem as the centre, or navel, of the Earth. See Antrim, *Routes and Realms*, p. 157, n. 36; A.J. Wensinck, *The Ideas of the Western Semites concerning the Navel of the Earth*, Johannes Müller, Amsterdam, 1916, pp. 21, 36.

26 See Ari Gordon, 'Sacred Orientation: The *Qibla* as Ritual, Metaphor, and Identity Marker in Early Islam', PhD dissertation, University of Pennsylvania, 2018.

Further reading

Abisaab, Rula, *Converting Persia: Religion and Power in the Safavid Empire*, I.B. Tauris, London, 2004.

Abu-ʿUksa, Wael, 'Lives of Frankish Princes from al-Ṣafadī's Biographical Dictionary, *al-Wāfī bil-Wafayāt*', *Mediterranean Historical Review*, vol. 32, no. 1, 2017, pp. 83–104.

Ahmad, S. Maqbul, 'Cartography of al-Sharīf al-Idrīsī', in J.B. Harley and David Woodward (eds), *History of Cartography*, vol. 2, book 1, *Cartography in the Traditional Islamic and South Asian Societies*, University of Chicago Press, Chicago, IL, 1992, pp. 156–74.

al-Idrīsī, Muḥammad ibn Muḥammad, *La première géographie de l'Occident*, trans. Amédée Jaubert, rev. Henri Bresc and Annliese Nef, Flammarion, Paris, 1999.

al-Idrīsī, Muḥammad ibn Muḥammad, *Nuzhat al-Mushtāq fī Ikhtirāq al-Āfāq: opus geographicum, sive 'Liber ad eorum delectationem qui terras peragrare studeant'*, 9 parts in 2 vols, ed. Alessio Bombaci, Umberto Rizzitano, Roberto Rubinacci and Laura Veccia Vaglieri, Istituto Universitario Orientale di Napoli/Istituto Italiano per il Medio ed Estremo Oriente, Naples, 1970–6.

Ali, Tariq, *A Sultan in Palermo*, Verso, London and New York, 2006.

al-Muqaddasī, *The Best Divisions for Knowledge of the Regions: A Translation of 'Aḥsan al-Taqāsīm fī Maʿrifat al-Aqālīm'*, trans. Basil Anthony Collins, reviewed Muhammad Hamid al-Tai, Garnet Publishing/ Centre for Muslim Contribution to Civilization, Reading, 1994.

Amara, Alloua, and Annlies Nef, 'Al-Idrisi et les Hammudides de Sicile: nouvelles donnees biographiques sur l'auteur du *Livre de Roger*', *Arabica*, vol. 48, 2001, pp. 121–7.

Antrim, Zayde, *Mapping the Middle East*, Reaktion Books, London, 2018.

Antrim, Zayde, *Routes and Realms: The Power of Place in the Early Islamic World*, Oxford University Press, Oxford, 2012.

Beeston, A.F.L., 'Idrisi's Account of the British Isles', *Bulletin of the School of Oriental and African Studies*, vol. 13, no. 2, 1950, pp. 265–80.

Benchekroun, Chafik, 'Requiem pour Ibn Hawqal: sur l'hypothèse de l'espion fatimide', *Journal Asiatique*, vol. 304, no. 2, 2016, pp. 193–211.

Blake, Stephen, *Time in Early Modern Islam: Calendar, Ceremony, and Chronology in the Safavid, Mughal, and Ottoman Empires*, Cambridge University Press, Cambridge, 2013.

Brotton, Jerry, *A History of the World in Twelve Maps*, Penguin Books, London, 2013.

Buondelmonti, Cristoforo, *Description of the Aegean and Other Islands*, ed. and trans. Evelyn Edson, Italica Press, New York, 2018.

Casale, Giancarlo, *The Ottoman Age of Exploration*, Oxford University Press, Oxford and New York, 2010.

Daftary, Farhad, *The Ismaʿilis: Their History and Doctrines*, 2nd edn, Cambridge University Press, Cambridge, 2007.

Ducène, Jean-Charles, *L'Europe et les géographes arabes du Moyen Age (IXe–XVe siècle): 'la grande terre' et ses peoples: conceptualisation d'un espace ethnique et politique*, CNRS, Paris, 2018.

Emiralioğlu, M. Pinar, *Geographical Knowledge and Imperial Culture in the Early Modern Ottoman Empire*, Ashgate, Burlington, VT, 2014.

Gautier-Dalché, Patrick, *La géographie de Ptolémée en Occident (IVe–XVIe siècle)*, Brepols, Turnhout, 2009.

Harley, J.B., and David Woodward (eds), *History of Cartography*, vol. 2, book 1, *Cartography in the Traditional Islamic and South Asian Societies*, University of Chicago Press, Chicago, IL, 1992.

Ibn al-Mujāwir, Yūsuf ibn Yaʿqūb, *A Traveller in Thirteenth-Century Arabia: Ibn al-Mujāwir's 'Tārīkh al-Mustabṣir'*, trans. G. Rex Smith, Routledge, London, 2017.

Kahlaoui, Tarek, *Creating the Mediterranean: Maps and the Islamic Imagination*, Brill, Leiden and Boston, 2018.

Kahlaoui, Tarek, 'The Maghrib's Medieval Mariners and Sea Maps', *Journal of Historical Sociology*, vol. 30, 2017, pp. 43–56.

Kennedy, E.S., and M.H. Kennedy, *Geographical Coordinates of Localities from Islamic Sources*, Institut für Geschichte der Arabisch-Islamischen Wissenschaften an der Johann Wolfgang Goethe-Universität, Frankfurt am Main, 1987.

King, David A., 'Astronomy in the Service of Islam', Lecture delivered at the al-Furqān Islamic Heritage Foundation, London, 7 March 2018, www.al-furqan.com/gallery/id/2668/filetype/video, accessed 18 February 2019.

King, David A., 'From Petra Back to Makka: From "Pibla" Back to Qibla' [Critique of Dan Gibson, *Early Islamic Qiblas: A Survey of Mosques Built between 1AH/622 CE and 263 AH/876 CE*], www.muslimheritage.com/article/from-petra-back-to-makka, accessed 18 February 2019.

King, David A., *In Synchrony with the Heavens: Studies in Astronomical Timekeeping and Instrumentation in Medieval Islamic Civilization*, 2 vols, Brill, Leiden, 2004.

King, David A., *World Maps for Finding the Direction and Distance of Mecca: Innovation and Tradition in Islamic Science*, al-Furqān Islamic Heritage Foundation/Brill, London, Leiden and Boston, 1999.

King, David A., and Richard P. Lorch, 'Qibla Charts, Qibla Maps, and Related Instruments', in J.B. Harley and David Woodward (eds), *History of Cartography*, vol. 2, book 1, *Cartography in the Traditional Islamic and South Asian Societies*, University of Chicago Press, Chicago, IL, 1992, pp. 189–205.

McIntosh, Gregory C., *The Piri Reis Map of 1513*, University of Georgia Press, Athens, GA, 2000.

Newman, Andrew J., *Safavid Iran: Rebirth of a Persian Empire*, I.B. Tauris, London, 2006.

Newman, Andrew J., 'Towards a Reconsideration of the "Isfahān School of Philosophy": Shaykh Bahāʾī and the Role of the Safawid ʿUlamāʾ', *Studia Iranica*, vol. 15, 1986, pp. 165–99.

ÖÎzen, Mine Esiner, *Pirî Reis and his Charts*, N. Refioğlu, Istanbul, 1998.

Pinto, Karen, *Medieval Islamic Maps: An Exploration*, University of Chicago Press, Chicago, IL, and London, 2016.

Pîrî Reis, *The Book of Bahriye*, ed. and trans. Bülent Özükan, Boyut Yayınları, Istanbul, [2013].

Porter, Venetia (ed.), *Hajj: Journey to the Heart of Islam*, British Museum Press, London, 2012.

Rapoport, Yossef, and Emilie Savage-Smith (eds and

trans), *An Eleventh-Century Egyptian Guide to the Universe: The 'Book of Curiosities'*, Islamic Philosophy, Theology and Science, Texts and Studies 87, Brill, Leiden, 2014.

Rapoport, Yossef, and Emilie Savage-Smith, *Lost Maps of the Caliphs: Drawing the World in Eleventh-Century Cairo*, University of Chicago Press, Chicago, IL; Bodleian Library, Oxford, 2018.

Rogers, J.M., 'Itineraries and Town Views in Ottoman Histories', in J.B. Harley and David Woodward (eds), *History of Cartography*, vol. 2, book 1, *Cartography in the Traditional Islamic and South Asian Societies*, University of Chicago Press, Chicago, IL, 1992, pp. 228–55.

Sarıcaoğlu, Fikret, *Pîrî Reîs'in dünya haritası 1513 = The World Map of Pîrî Reîs 1513*, T.C. Kültür ve Turizm Bakanlığı, Ankara, 2014.

Savage-Smith, Emilie. 'Celestial Mapping', in J.B. Harley and David Woodward (eds), *History of Cartography*, vol. 2, book 1, *Cartography in the Traditional Islamic and South Asian Societies*, University of Chicago Press, Chicago, IL, 1992, pp. 12–70.

Savage-Smith, Emilie, 'Memory and Maps', in Farhad Daftary and Josef W. Meri (eds), *Culture and Memory in Medieval Islam: Essays in Honour of Wilferd Madelung*, I.B. Tauris, London, 2003, pp. 109–27, figs 1–4.

Sezgin, Fuat, *Mathematical Geography and Cartography in Islam and their Continuation in the Occident*, 3 vols, English translation of vols 10–12 of *Geschichte des Arabischen Schrifttums*, Institute for the History of Arabic-Islamic Science, Frankfurt am Main, 2000–7.

Soucek, Svat, 'Islamic Charting in the Mediterranean', in J.B. Harley and David Woodward (eds), *History of Cartography*, vol. 2, book 1, *Cartography in the Traditional Islamic and South Asian Societies*, University of Chicago Press, Chicago, IL, 1992, pp. 263–87.

Soucek, Svat, *Piri Reis and Turkish Mapmaking after Columbus*, Oxford University Press, Oxford, 1996.

Tibbetts, Gerald R., 'The Balkhī School of Geographers', in J.B. Harley and David Woodward (eds), *History of Cartography*, vol. 2, book 1, *Cartography in the Traditional Islamic and South Asian Societies*, University of Chicago Press, Chicago, IL, 1992, pp. 108–36.

Tibbetts, Gerald R., 'The Beginnings of a Cartographic Tradition', in J.B. Harley and David Woodward (eds), *History of Cartography*, vol. 2, book 1, *Cartography in the Traditional Islamic and South Asian Societies*, University of Chicago Press, Chicago, IL, 1992, pp. 90–107.

Tibbetts, Gerald R., 'Later Cartographic Developments', in J.B. Harley and David Woodward (eds), *History of Cartography*, vol. 2, book 1, *Cartography in the Traditional Islamic and South Asian Societies*, University of Chicago Press, Chicago, IL, 1992, pp. 137–55.

Winterburn, Emily. 'Using an Astrolabe', *Muslim Heritage*, http://muslimheritage.com/article/using-astrolabe, accessed 18 February 2019.

Zadeh, Travis, 'Of Mummies, Poets, and Water Nymphs: Tracing the Codicological Limits of Ibn Khurradādhbih's Geography', in Monique Bernards (ed.), *ʿAbbasid Studies IV: Occasional Papers of the School of ʿAbbasid Studies (Leuven, July 5–July 9, 2010)*, Gibb Memorial Trust, Warminster, 2013, pp. 8–75.

Index

Acknowledgements

Samuel Fanous, Head of Publishing at the Bodleian Library, came up with the idea for this book nearly a decade ago, and has been gently but persistently nudging me to bring it into existence. I am very grateful to him for doing so. It has been a real pleasure working with my editors at Bodleian Library Publishing: Janet Phillips, Deborah Susman and especially Leanda Shrimpton who tirelessly matched my text with the perfect image.

In the course of writing this book, I benefited much from the advice and expertise of my fellow historians of cartography. Zayde Antrim generously shared her comprehensive knowledge of the Islamic mapping tradition, and I owe much to her conceptual framework. I am very grateful to Nadja Danilenko, who went through a draft of Chapter 2, and made numerous suggestions and corrections. Ari Gordon shared with me his thought-provoking research on the *qibla* as metaphor and identity marker. Omar Anchassi opened my eyes to the medieval Islamic adherents of the idea that the Earth was flat. My map history colleagues and friends at Queen Mary University of London, Jerry Brotton and Alfred Hiatt, have been extremely supportive. At Oxford, I also owe special thanks to Silke Ackermann, director of the History of Science Museum, and to Nick Millea, Map Librarian at the Bodleian, for guiding me through the collections.

Above all, I am deeply indebted to the generosity and immense knowledge of Emilie Savage-Smith, legendary Professor of the History of Islamic Science at Oxford, who introduced me to the field of Islamic cartography twenty years ago, and has guided me ever since. Her insistence on approaching medieval Islamic science as both a science and an art, and on placing it in its historical context, shaped the approach taken in this book. I have been so very fortunate to have her as mentor, collaborator and friend.

First published in 2020 by the Bodleian Library
Broad Street, Oxford OX1 3BG
www.bodleianshop.co.uk

ISBN: 978 1 85124 492 8

Jacket design by Dot Little at the Bodleian Library
Designed and typeset by Ocky Murray in 11pt Gentium
Printed and bound by Printer Trento S.r.l on 150gsm Gardamatt paper

FSC www.fsc.org MIX Paper from responsible sources FSC® C015829

British Library Catalogue in Publishing Data
A CIP record of this publication is available from the British Library